COSMIC ADVENTURE

ALSO BY BOB BERMAN

Secrets of the Night Sky: The Most Amazing Things in the Universe You Can See with the Naked Eye

COSMIC ADVENTURE

A Renegade Astronomer's Guide to Our World and Beyond

BOB BERMAN

Illustrations by Alan McKnight

William Morrow and Company, Inc.
New York

It is the policy of William Morrow and Company, Inc., and its imprints and affiliates, recognizing the importance of preserving what has been written, to print the books we publish on acid-free paper, and we exert our best efforts to that end.

Library of Congress Cataloging-in-Publication Data

Berman, Bob.
Cosmic adventure : a renegade astronomer's guide to our world and beyond / Bob Berman ; illustrations by Alan McKnight.
p. cm.
ISBN 0-688-14495-0
1. Serendipity in science. 2. Science—Philosophy. 3. Astronomy—Philosophy. I. McKnight, Alan. II. Title.
Q172.5.S47B47 1998
500—dc21 98-17819
CIP

Printed in the United States of America

First Edition

1 2 3 4 5 6 7 8 9 10

BOOK DESIGN BY RENATO STANISIC

www.williammorrow.com

To my beautiful, playful eight-year-old daughter, Anjali—the sun around which my life revolves . . .

And to her mother, Rita, whose extraordinary character and happy disposition have been an inspiration

Acknowledgments

My eternal gratitude to Paula Dunn for tireless edits and invaluable suggestions. Thanks also to Jane and Larry Weinberg for their encouragement, Steve Charney for his astute comments, and Dr. Christopher Coulon for his observations about water, which appear in the chapter "Everyday Alchemy."

Selected phrases and even whole paragraphs of three chapters in this volume, especially "Bohdan's Bursters," originally appeared in my articles published in *Astronomy* magazine, *Discover* magazine, and *Sky and Telescope* magazine between 1994 and 1998.

Contents

Introduction

The universe hardly needs a public relations "rep," but by sheer good luck or divine design, that has been my job description since 1974. That's when my first astronomy column was launched; it was written for *Woodstock Times,* the largest weekly newspaper in the upstate New York area I call home—and I'm still meeting that deadline.

The sky has always been my passion. Before I was thirteen, I'd read every astronomy book in the local library; while friends were absorbing baseball stats, I memorized the names and distances of every star in the heavens; and in college, I took every astronomy course offered. Only my involvement with skydiving and overseas motorcycle touring prevented an astronomical case of nerdship.

Luckily, I've been able to combine my sky fixation with a parallel love of writing in a career that lets me present the facts, discoveries, concepts, and awesome wonders of the universe the way people (myself included) really like to learn about them—and that is with exuberance and humor rather than the needlessly pedantic drone of "cultural enrichment" or "science education." Years of lecturing have taught me that most people have no desire to fool with charts or learn the fainter constellations; they appreciate when astronomy's newest fascinating facts are served with playfulness, or maybe even a philosophical outlook.

This book allows me to explore topics beyond the thematic limitations of my *Discover* magazine column or my first book, *Secrets of the Night Sky,* which cover amazing, little known, and sometimes funny aspects of celestial objects. For in those arenas I am unable to include an underlying realm—astounding, seldom-examined facets of the world around us and the universe beyond. Here, I hope, the reader will join me in venturing to the heart of things fabulous, frivolous, and even hilarious. We'll probe the quirky nature of some basic physical realities (for example, why water, the most common compound in the cosmos, isn't a gas at room temperature like other molecules of its size and weight) and have fun orienting ourselves to celestial-language oddities ("to orient yourself" comes from the old belief that you'd be headed correctly if you knew which way East—the Orient—lay).

Among other challenges, readers will address the link between contemporary ideas of cosmology and the limitations of the human brain's logic system, a linguistic handicap in the way of solving such issues as "What lies outside the universe?"

Here I can finally present astronomy in a broader context: What was the single most astonishing discovery of all time? Why is there more cloudy weather around the time of full moon? What, realistically, would an alien invasion *really* be like? What were the biggest cosmic goof-ups, the greatest space disasters?

This book allows us to probe such odd alleyways as the official international organization that labels the universe's contents—with such disparate names as Puck and Centaurus A—and the story behind that company in Illinois that names stars after people, for a price.

Here we look at the sacred chestnuts and the cosmological clichés, and explore everything from bizarre Einstein lenses that warp galaxies like funhouse mirrors, to the increasingly persuasive nonexistence of time, to myths and little-known facts about the extreme edges of the cosmos.

To a public that is generally clueless about astronomy (how many know that the moon crosses the sky from left to right?), I wanted to impart fascinating data such as the fastest spins of diamond-hard pulsars; what it would really take to terraform Mars; what is the only truly original concept in cosmology; and how anyone can quickly figure the distance to the horizon when flying in a commercial jet.

It all makes this book a motley stew that has been prepared and seasoned not only for the intelligent layperson with no astronomy background, but certainly also for the celestial devotee who is overdue for a breeze to blow in fresh ways of viewing old celestial friends.

For most people, a particular scent can evoke a memory. For me, heavenly bodies sometimes flash a personal earthly connection. Since it's almost impossible to separate the astronomy part of my life from the rest of it, these essays include some personal journeys and reflections; I've also confided accounts of a few of the many strange people, events, and experiences that sprang to mind while researching and writing this book.

Neither my talented editors nor I could devise some logical order to the following chapters. Hubris, of course, but I like to think that reflects on a microscale the wild abandon seen orchestrated in the layout of the galaxies and constellations.

And that is why our topics meander from compulsions to time running backward to the discovery of Uranus to the explosions of the *Challenger* and the *Hindenburg*. The only link is my honest desire to have readers accompany me on quests to stretch the mind, challenge prevailing concepts, and explore ideas (from idiotic to inspired) about the very nature of the cosmos and the capabilities of human intellect. It's a roller-coaster ride.

I hope you enjoy the turns and surprises as much as I have.

COSMIC ADVENTURE

Heartbreak from the Comet Cloud

Either life is always perfect and this flawlessness is cloaked by our ignorance, or else frightening snakelike patterns slither menacingly across a cosmos whose motif is accidents and heartbreak.

Buddhists and Hindus have it both ways. Their pundits declare that a fascinating bit of theater is at play, creating the ups and downs of life—a seductively realistic chimera—while a deeper meaning lurks in everyone's underlying storybook. Blown opportunities coexist with a perfection worthy of a finely crafted novel. Wheels within wheels.

Granted, it's hard to fathom any "deeper meaning" when fifteen years' work vanishes as an exquisitely instrumented unmanned spacecraft explodes on liftoff. Or find consolation for the tenacious comet hunter who finally discovers what becomes the brightest comet of the century—but finds it six hours after someone else. So it is another whose name gets blazoned across the heavens, who gains instant celebrity, lands prestigious appointments, and rides off into the moonset.

The solar system, too, is full of losers. Venus is just 30 percent closer to the sun than Earth, yet that's enough to make it the most stifling hellhole in the known universe. Going the other way, a fifth planet, beyond Mars, never quite formed because of insistent gravitational meddling from outer planets. What lesson is there from one of the solar system's children being stillborn?

The most gifted artist I ever met spent an entire year on his magnum opus. Oils on which he invests a mere two weeks look like museum treasures, so I can only wonder about the glorious masterpiece forged by that year's dedication. I'll never know, because when the painting was finally completed, he wrapped and carefully tied it to the rooftop luggage area of the intercity bus on which he was traveling in Asia. But upon reaching his destination, he discovered that the painting was gone. The twine had snapped!

Like a crazed, disconsolate wanderer, he pathetically set out on foot to retrace the bus route. He walked the entire sixty miles, examining both sides of the rural road, stopping at every village to offer a generous reward. All for nothing.

What's the purpose of that loss?

During the Asian total eclipse of 1981, a group of American astronomers had been invited by Soviet authorities to observe from an ideal site on an island in Lake Baikal. On the morning of the eclipse, however, two astronomers overslept and missed the ferry. The pair had to content themselves with viewing the event from their hotel rooftop (where they actually saw it quite well). Meanwhile, a stationary cloud formed over the lake, and nobody at the "perfect" site saw a thing! The late sleepers were the only Americans to observe the great eclipse.

What does that tell us?

Are events in our lives—or in the cosmos itself—random patterns shifting with blind abandon, or is there some greater design too intricate to be perceived by our limited vision? From our egocentric point of view, we're torn and tortured over disappointments no matter how we try to rationalize and shrug them off with "Oh, well, it was meant to be!" Many of us always want to coerce improvement in our circumstances and yet, when pressed, admit that we could not have written a better scenario for our lives than the way things spontaneously unfolded. Sad interludes often lead to happy finales. On the coin's flip side, our hearts' desires may deteriorate with time, grand expectations evaporate. Somewhere, a mother's sunflower eyes beam at the first smile of her infant, destined someday to hold up a convenience store. Darkness before dawn. Azure sky followed by hurricane. Destructive novas generating starbirth.

Opposing layers of agony and ecstasy, splendor and pettiness, conform to the cosmic infatuation with alternating rhythms. Our very thoughts along these lines, racing at 250 miles per hour through impossibly labyrinthine neural pathways, may display a consonance with the universe's operating system. A trillion brain cells are nourished and maintained so that a few millivolts of electricity can snake through their numbing complexity—all so that a teenager can apply lipstick! It's as if the energy of the world's waterfalls were focused into one ultrapowerful, galaxy-spanning broadcast of *Wheel of Fortune*. Grandeur harnessed for triviality.

Like copper wire within an insulator, the carrying cables of life's intensity often seem constrained by petty intent. Speaking requires the use of seventy-two muscles, each sustained by the blood coursing through countless capillaries. Yet the typical end result of this architectural triumph is a cliché. *Check out that convertible, man. Awesome.*

Once this design of alternating forces is glimpsed, it's easier to see banality in a new light, as merely a segment of equal and opposite formations, an amusing yin of a complete yin/yang. Confined to our lives' microscopic perspective, we see only the comedy or the tragedy; a larger, broader view would reveal the stasis, the equilibrium, the next stratum—if it weren't *so* profound as to escape detection altogether.

Maybe the "music of the spheres" is a jazz riff, improvised as

Stardust has worked its way through evolution's maze to produce a bewildering species that spends far more on cosmetics than on science education.

it goes along. Or perhaps it's a symphony, carefully crafted of disparate elements heading toward harmonies, crescendos, and finales so exquisite that an immortal audience would be moved to tears.

So which is it—luck, or a fabulous script? And if the latter, is this how the universe works at large? Are star clusters evolving toward some Grand End?

It's no longer such a far-out idea. Leaving aside metaphysical considerations, we already know that each biological cell is designed to work for the overarching benefit of the individual animal or plant. And we observe biomes—large, harmonious communities of plants, microbes, and animals—that have a symbiotic relationship of codependency. In short, individuals live within an intelligently designed matrix. (We need not decide whether such intelligence evolves as it goes along or is dictated by a deeper underlying faculty—any more than an astute question requires proof of origin.)

But does this process stop at the biome level? James Lovelock's Gaia hypothesis (which is really a restatement of ancient Eastern thought) says that Earth itself is an intelligent entity. That is, all its myriad biological systems fit together in a sage and self-regulating fashion. Throw off the global carbon dioxide balance, and oceanic plankton will multiply to absorb it. By this reasoning, humans are allowed only so much meddling in the Earth's ecosystem. If our actions become excessive, any one of several natural mechanisms will arise to take care of it, or us.

And why not? Humans, after all, sprang naturally from Earth. Our large, technology-creating brains, as convoluted as our rationalizations, do not stand apart from the biosphere but arose from it like seeds within an apple. If nature is thoughtful rather than merely intelligent, then we are myrmidons in an ongoing, never-ending project. Our aggression may be as preprogrammed as a computer booting up. If we eventually commit the ultimate "blunder" and engage in nuclear war, then it would actually be no error at all, but what we were designed to accomplish all along.

If something is infinite, then no finite amount of screwing up can do any real damage. Perhaps humans were *intended* to mine uranium and release radiation, whose effects are ultimately good—if the goal is to accelerate evolution. Most radiation-induced genetic changes are unfavorable or even fatal. But bad mutations die out while beneficial ones thrive. A nuclear war would annihilate much, but it could not destroy all life. Earth's entire biological system would explore new pathways, enjoying a hundred million years' worth of evolution in a few brief millennia. Earth's biosphere would take the *Reader's Digest* condensed route to the next level of its collective destiny. The change, the adventure nature always seems eager to undertake would be sped up. And poetic justice: this impatient, hurry-up species, the humans, unleashing a hurry-up potion upon the planet and upon ourselves.

Not that nature *always* enjoys fast action. Cockroaches have remained unchanged for 250 million years. Impervious to radiation, they're not likely to go along with the plan and metamorphose into anything else. (A good thing. It's horrible to imagine a future human generation trying to exterminate an improved roach.) Personally, I will do everything an individual can do to help prevent nuclear accident. But if the Big One comes anyway, I hope I'll remember these thoughts in my final moments, to salute the stunning, here-comes-a-new-Earth mushroom cloud with a toast of: Excelsior!

What of the larger universe? Is the solar system evolving? And can we ever discover the master plan, the underlying scenario of the stars? With the recent finding that Mars may have had microbial life billions of years ago, it's tempting to visualize the planetary arena as a kind of marathon race. Mars, smaller and farther from the sun, probably cooled first and could well have offered ripe conditions for life several hundred million years before Earth. Once established, microbial life could have been blasted off by the meteor impacts that were then common, and some may have fallen to Earth to establish terrestrial colonies. By this reasoning, we are all Martians! Except the building blocks of life, the amino acids, may have originally come to Mars from yet somewhere else!

Six billion years hence, when the sun has collapsed in its old age into a feeble white dwarf—the shrunken end point of all solar-type stars—it may finally be Venus' turn to live up to its name and become paradisiacal. Like a fetus with a lifetime's destiny ahead, our human era may represent just the early springtime of the solar system. As the second millennium begins, we glimpse in our own lifetimes but the snap of a finger in an unfolding epic that may embrace far more than the biography of our planet alone.

And how about larger plans—for the galaxy, or our home cluster of galaxies, or the entire observable universe? Was the Big Bang just the labor pains at the birth of one fantastically complex offspring in an eventual cosmic litter of trillions—whose destinies are part of a cosmic scheme unfathomable by any science?

An alternative conclusion, of course, is the traditional view of dumb matter and unfocused energy obeying blind laws of gravity and chance. No plans at all, and no theater. The pinball machine without the player. Intricate blueprints springing from random events. The monkey and typewriter thing. Put sufficient quarks into the blender, press "liquefy" for enough aeons, and out pops Vivaldi and golden retrievers.

By this account, there are indeed accidents and tragedies, and life's hold on Earth is fragile, not self-regulating. It's hardball and it's for keeps. No grand scheme at all, except what we humans can come up with. And our own strategy had better be good, because that's All There Is.

From our vantage point on Heartbreak Hill, the Milky Way may as well be the Wild West. Supernovas *ka-blam* like Russian nuclear power plants, taking out any unfortunate alien life on nearby planets. No design here, just hydrogen doing its one-step.

The problem, of course, is that there is no handbook to guide us in deciding which operating system governs the cosmos. It's a matter of outlook. My own bias should be apparent: If this is not a shadow play being performed, then we've forgotten how to recognize one. But I doubt I can influence anyone who isn't already at least in the undecided corner.

Some of the most brilliant men and women I know, people who aced the Miller analogy test given to graduate students, see the universe as a pool table where imbecilic billiard balls carom. They imagine that their own smarts somehow sprang from witless bits of atomic flotsam, like pansies from horse manure. I could accommodate them by providing a compelling description of how life could self-arise from random events. But you still don't have the recipe for consciousness, for awareness itself. *That* remains the stopper, the mystery as wonderful and fascinating as when an infant first sees her fingers.

Justifying either view doesn't make much sense, because one or the other will ring true depending on one's philosophy. A positive spin on nuclear war will seem a preposterous example of rationalization run amok, especially if the reader sees death as unfortunate and the deaths of entire planets as the quintessential bad day at the races.

But life is so full of seemingly unlucky events proving otherwise and vice versa that it's shaky to credit our own limited judgment of anything whatsoever.

There's an illustrative story I once heard in the East, of a poor but wise farmer who was very aware that things are often not as they first seem. One day his only horse ran away, and sympathetic neighbors cried, "How terrible." But the farmer just shrugged. "Maybe."

The next day the horse returned, leading ten wild horses. Suddenly the farmer was rich! The neighbors gathered around and shouted, "How wonderful!" The farmer said, "Maybe."

Later, his only son, while trying to break in one of the horses, was thrown, and his leg horribly broken. The neighbors said, "How terrible." The farmer said, "Maybe."

The next day, army officials came to draft the boy for the war but, seeing the mangled limb, left without taking him. Neighbors said, "How fortunate!" The farmer said, "Maybe."

On it went. The point, of course, is that our immediate assessment of affairs cannot be trusted, because we don't know what lies ahead. Whether or not a Grand Plan is at play, unseen twists farther downstream make current events merely a segment of a process that is, at minimum, rather mysterious.

That's why it is a little surprising that some astronomers, who postulate that our sun is a member of a binary system, name the companion Nemesis. This ominous appellation caught on because, when the dark star approaches us in its huge looping orbit, every 80 million years or so, its gravitational influence roils the halo of comets that make up the distant Oort cloud. Numerous airport-sized cannonballs are pitched inward from this reservoir of icy rocks, assaulting the solar system like a Saturday-night riot in an inner city.

One of these clobbered Earth 65 million years ago and eighty-sixed the dinosaurs. The geologic record shows that such semiperiodic violence is our destiny time and again.

But why call it Nemesis if the death of the raptors suddenly allowed mammals to flourish and paved the way for humans? It might be more fitting to name the companion star Serendipity. The reason for the forbidding label, I think, is that a dolorous outlook

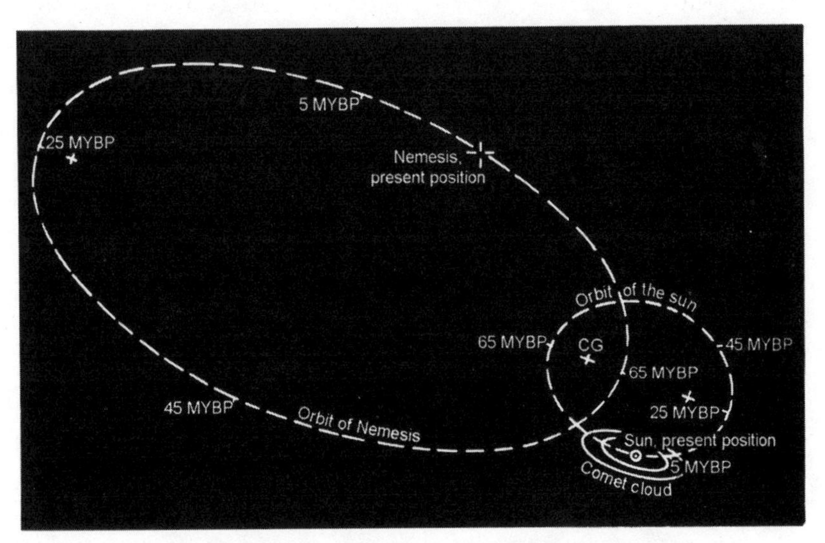

Our sun and its hypothetical companion, Nemesis, follow elliptical paths around their common center of gravity, CG. Most of the time, Nemesis is too distant to affect our solar system, but every 80 million years or so, it swings close enough to disrupt the comet cloud. Positions of the sun and Nemesis are labeled MYBP for "million years before present."

is more acceptably in sync with the "stupid universe" model. The notion that "we" are the first single-celled organisms, the dinosaurs, and the early mammals, that "we"—meaning consciousness—have explored Earth through many different eyes, and that changes are good, even fun, is too all-embracing.

Science is supposed to be dispassionate, but when it must take a slant, the tacitly correct viewpoint is one that is rigidly parochial. Textbooks have no choice but to define "we" as humans alone and to see major alterations in earthly reality as grim, scary accidents. That last dinosaur-demolishing event gets the stamp of approval only because it brought us here as abruptly and unexpectedly as a mouse darting from under the fridge.

Yet even with this beneficent turn of events, this reason to celebrate the giant gong that signaled the end of the Cretaceous Period, we *still* call the star Nemesis.

Go figure.

Original Thinking

"Did you lure the raccoon with Ritz crackers?"

It certainly was a reasonable question. The same raccoon had kept breaking and entering through the cat door night after night. After days of awakening to a kitchen scene from Dante, littered with abstract debris from the overturned garbage can, I set a Havahart trap baited with crackers. It was the best I could do since I had no chocolate, which supposedly they cannot resist.

But morning dawned happily. The racconian nightmare was over. Caged and anxious, the marauder and his half-eaten bait awaited transplantation ten miles down the road, to be released unharmed near my ex-wife's house. And this is when my friend arrived, noticed the Nabisco detritus, and popped the odd question.

We've all heard the cliché about clichés, that nobody says anything original; everything that *can* be said has already been expressed. But I was quite sure I'd never been asked about Ritz crackers and raccoons.

So it is with new ideas. Mostly there aren't any. Starting from cosmic origins and the Big Bang theory—which essentially maintains that the entire universe traces its origins to a precise natal instant—celestial ideas are mostly extrapolations from what we see here on Earth, where each opossum or French pastry arises at a fixed moment in time. It stands opposed to, say, dark matter and time itself, whose blurry roots date back at least 11 billion years but are sufficiently unknown to permit the notion that they may have *always* existed, or perhaps even squeezed here from another dimension, possibly with forms altered from an earlier incarnation.

That's why when a truly new thought comes along, we recognize and honor it. It's not easy to recall ideas that are more than mere carryovers from other disciplines. One of them, perhaps, was the Hundredth Monkey concept of the 1970's.

The scenario: A researcher observing monkeys on a particular island noticed that one started washing its food before eating it, to remove the sand. Soon others were doing the same—behavior never before seen in this type of monkey. A sort of evolutionary action was apparently occurring.

Now the spooky part. When it had reached the point where a great number of monkeys routinely acted in this manner, researchers suddenly started seeing the same behavior among this species of monkey in other parts of the world. The conclusion was startling: When some large number of creatures begin thinking a particular way, the situation reaches a sort of critical mass and the idea pops into the minds of *all* those animals simultaneously, no matter where in the world they may be.

It was a novel biological idea, even if the use of the term *critical mass* does suggest an analogy to another phenomenon—nuclear fission and the atomic bomb, where things suddenly and explosively get out of hand once a pivotal point is reached. And while New Age in flavor, it didn't seem totally preposterous. We've all seen large flocks of birds or schools of fish make seemingly simultaneous turns, and other life-forms do indeed appear mentally connected, as if an unseen neural thread binds them even when separated by large distances. It's the ESP thing, never quite disposed, never really laid to rest by science.

Ken Keyes grabbed and further New-Aged the idea in his popular book *The Hundredth Monkey*. Keyes proposed that if enough people embrace and participate in the peace and environmental movement, it will suddenly "catch fire" and become the established paradigm of the human race.

Well, this is a wonderful thought, and it may even happen. But meanwhile it turned out that the original story was apocryphal. It seems there never was a researcher who noticed increasing numbers of monkeys washing fruit. Indeed, from time to time monkeys have always washed fruit.

The point is that a novel idea can be compelling enough to have its own life *even if it isn't true*. Like a computer virus, it spreads readily from brain to brain precisely because it is catchy. No, more than that—catching; contagious. In a very real sense, original, easily grasped concepts are self-replicating. They are like the latest punchy, topical jokes. Once released into the environment, they reproduce themselves seemingly beyond human control.

The same appears true of astronomical and cosmological ideas that have reached the general media to become the paradigms that have self-replicated until widely accepted as givens. Or, having failed to flourish, despite being valid, have absented themselves from our collective model of the cosmos.

The cosmological bombshells of the late twentieth century are good starting points. Item one: the age paradox. Its background is simple enough: The expansion rate of the universe lays down definite restrictions on how old the cosmos may be. In 1996, new evidence from the Hubble Space Telescope showed that this is somewhere between 11 billion and 14 billion years. The problem was that most of the universe's stars seem older than this. (The paradox has since been largely, but not fully, resolved.) Moreover, the large-scale structure of galaxy clusters obviously needed far more time to have evolved into the colossal formations we observe, and the amount of helium in the universe, according to some respected observations, fails to match the critical amounts predicted by the Big Bang theory. Translation: The Big Bang had to have happened, but the Big Bang couldn't have happened.

Such contradictions make us uncomfortable: We sorely want the universe to make sense (even if the universe is composed of subatomic particles that follow quantum laws that very definitely do *not* make sense).

Item two: big bubbles. Absolutely nobody anticipated the shocking finding of the 1980's that on large scales the universe is constructed like a sponge. Huge voids are surrounded by (defined by, actually) curving sheets of galaxy clusters. How did *that* arise?

Big bubbles and the paradoxes of the Big Bang are stupendous questions in every sense. Yet these ideas never made the cocktail-conversation circuit despite being easy to grasp and having good potential for intelligent discussion. The reason, I think, is that they didn't meet the self-replicating criterion of being analogous to *other* concepts. In other words, there has to be a thread connecting it with something already intimately well known.

Planets, by contrast, do meet this standard. The media quickly grasped that there was plenty of "ink" in the notion of "new planets around other stars." No matter that the discoveries involved technical intricacies (like Doppler shifts in the spectral lines of pulsars, or minute irregularities in the motions of objects like 51 Pegasus) that are utterly unfamiliar to the public. The idea caught on because *planet* is one of the few concepts that has a secure home within everyone's brain. Add the word *new*, and we have a hundredth-monkeylike catchphrase: *new planet. New planet found around a star.* No word had more than two syllables. It made *Nightline* and prime-time news shows.

I'd love to see things that make *no sense at all* reach the mass media. Whoever threw the first banana cream pie in someone's face, or wore curlers, or kept a mouse as a pet, probably was a lunatic. But once such unlikely notions manage some critical number of repetitions to travel a well-worn behavioral or conceptual road, they become more than merely acceptable—in some cases their newly gained legitimacy acquires an aura of truth or becomes some sort of functional standard. This is reflected in the don't-rock-the-boat stance frequently taken by peer reviewers; in turn these reviewers influence the path researchers are encouraged to follow and, ultimately, what all of us know about the universe.

Take the well-known *red shift.* Here we have an archetypical catchphrase. Two easy words, one of which is a color. Somehow, a two-word expression containing a color is practically guaranteed to stick in the public mind. *White dwarf. Red giant. Black hole.* Despite very low general awareness of basic astronomy, these phrases manage to catch on and be known by all.

So *red shift* is "in" despite having a definition that not one in a thousand people could manage: that spectroscopically observed absorption or emission lines from a light-emitting object shift toward the longer (redder) wavelengths when that object is receding, to a degree proportional to its velocity. Since distant objects show greater red shifts, it means that the whole universe is expanding.

So far, so good. But during the 1970's and 1980's, a vocal minority of astrophysicists, led by the well-respected Halton Arp, presented repeated evidence of physically linked galaxies (for example, with tendrils of gas between them) in which members displayed very different red shifts. In short, something else might be causing the red shift. If red shift equals distance, and if galaxies in a single group show different red shifts but clearly sit at the same distance from us, then something strange and unknown is afoot.

But this ongoing controversy barely made it into the journals, let alone the popular literature. Our present point is not whether red shift really is a dependable indicator of distance; in truth it almost certainly *is* trustworthy. The issue here is that red shift was okay. Much less okay is paradox or discomfiture.

If all physical laws either followed, or failed to follow, modern *Homo sapiens*' logic system, we'd enjoy a smooth philosophical ride. Our models of reality would be straightforward. If it's all consistently rational, then every question can be solved by math, science, or sufficient deep thought. If irrational, then we can simply laugh, drink our wine, and accept that life is an insuperable mystery. No problem either way.

Instead—and this is new in human thinking—logic works to perfection on some levels, while in other cases it seems to have no value at all. For example, we can predict eclipses to the minute because Newton's laws are obeyed by every dust mote out to the ends of space and time. But look at the electrons within that mote and it's a different story: Each behaves in antilogical, unpredictable ways; only large numbers of electrons can be understood as a group, and then only in terms of probability.

Suppose we construct a measuring device known to bias the properties of half the electrons passing through it (a property can be the spin around any of several axes, for example). If these electrons do not pass through the device, their properties stay the same; if they do pass through, they are changed in a particular way. But when we force them through a series of such devices, some characteristics that have nothing to do with these devices change, while others do not. Here's the point: The paths that some electrons have taken reveal that they have neither passed through the detectors nor *not* passed through. They have done something else. Something beyond our comprehension.

No one has the slightest idea what is going on, or how subatomic particles can perform these impossible feats of antilogic. That they do is duly noted, given a name (the electrons performing mind-boggling actions are said to be in *superposition*), and even exploited for our purposes or inventions. But naming these actions does not mean we can explain them. We cannot.

The fact that "impossible" things happen faithfully enough to be scientifically predicted is novel; it's a concept that first arose in the twentieth century. It is, in a way, new and catchy. But these ideas, as well, fail to make it into the mainstream hundredth-monkey mind-set. Not for lack of glitter and intrigue; they have plenty of both. But because, I think, they lack the familiar latch that could attach them to something else in the mind's experience.

So novelty isn't enough. Insight and profundity aren't enough. Originality must be packaged with a hook, like a cell's sticky protein coat, in order for the new idea to become infectious. A functional hook could be a short and catchy title or, better still, a commonality with older concepts so that it doesn't strike us as altogether strange. Like a mutating bacterium, it mustn't be an unrelated species, a complete alien.

This is why the Big Bang and other cosmological concepts tend to evolve in stages instead of with leaps of insight stemming from truly original thinking. Although Einstein's special relativity seemed explosively creative to the public, it was actually a piecemeal bit of evolution that built on the revelations of earlier physicists like Hendrik Lorentz and George FitzGerald. His general relativity, however, was a truly quantum jump. Most comparable leaps since then have not been lightning bolts of unfettered inspiration, but rather discoveries that thrust themselves upon unsuspecting investigators.

But such is the way of things. We'll rarely go wrong if we treat skeptically the leading catchy-named cosmological Top 40 item; the odds of its validity, like those of a Derby favorite, have been prejudiced by its popularity. Meanwhile, perhaps our collective thinking will evolve to the critical mass of enjoying rather than dodging paradox. Just as the next racconian generation, sired by previously captured and released parents, may come to favor the hors d'oeuvre cracker.

And the hundredth (like the apocryphal monkey) may, along with all his brethren everywhere, learn to put the lid back on the garbage can.

Cosmic Blunders

Our human universe, perhaps reflecting the larger one beyond, teems with blunders that are often more interesting—and always more enlightening—than when things go as planned.

In 1962, the spacecraft *Mariner 1,* bound for Venus, had to be destroyed because it strayed off course—a goof caused by the omission of a hyphen in the craft's computer programming. This single punctuation mark cost taxpayers $18 million.

Going outward in the opposite direction, the Russian spacecraft *Phobos,* en route to the Red Planet in 1994, blew up at the command of a programmer who sent a seriously flawed computer message. Obviously, the old saying about needing a computer to *really* screw things up rings just as true beyond Earth's atmosphere.

At your next party, go around the room and ask everyone: What was the greatest foul-up of your career? Those in the medical profession may be reluctant to answer, considering their possible involvement in causing unnecessary pain or premature death. Understandably, most others would rather not confess to gross negligence. But in a circle of close friends (and if you, yourself, start the ball rolling by confessing first), a few may be coaxed into admitting their most egregious mistakes.

Why perform such an exercise? Because the results are more than fascinating; they are revelatory. For there may be only one way to do something right, but there are an almost infinite number of ways to mess things up. That potential for variety, for the unexpected, provides the engrossing backdrop for life's tragicomedies and is especially riveting when it comes to the sciences.

Okay, I'll go first.

When I was in my twenties, after building my home more or less single-handedly, hubris led me to think that I had suddenly become a skilled carpenter—and electrician and plumber as well. Hadn't I wired up the house from scratch, and didn't it pass inspection?

Then, upon admiring my finished house, a friend who planned to open a restaurant in my hometown asked if I would do the construction.

Now I was a professional! I had found the way to supplement the meager financial rewards of astronomy: I would make my fortune as a contractor by day and do my starstuff by night. I hired helpers and thus began what became a ten-year foray into the construction business.

Incredibly enough, I had just two unhappy clients out of hundreds and never lacked for work just from word of mouth. Only when my parallel astronomy career as teacher, columnist, and author dominated my time did I put down tools forever. But in that decade, boy, did I screw up. More: Stories confided to me by other contractors revealed that mistakes were not only endemic but usually covered up. The homeowner or customer simply never knew.

But it wasn't others, but I, personally, who felled a tree that crashed into power lines and darkened a neighborhood for hours. It was I who punctured an electric line while drilling through a floor joist, plunging the crowded store above me into blackness.

It was my "expert" plumber who attached the hot-water line

to a toilet and then did not catch the error until the family moved in. Heated, steamy water in the bowl—can you imagine the owner's expression when his derriere encountered that unexpected and unsolicited luxury?

If something as low-tech as home-building can produce such a cornucopia of blunders, then what of more intricate endeavors? We've all seen films of bridges swaying and collapsing, of experimental planes that were dangerously unstable, of weaknesses in every area of technology. Even astronomers have been crushed by their own observatory domes. (Yet we are to believe that the world's nuclear arsenals, now largely controlled by computer, are totally fail-safe. Cross your fingers.)

Goof-ups in space science are usually costly, often spectacular, and always instructive. The archetypical botch-job, the event listed as an example when you look up *blunder,* was the 1990 fiasco surrounding the Hubble Space Telescope, whose 94-inch mirror was the most precisely polished ever made—except that it was cut to the wrong shape.

Never before had so many involved in the space sciences laughed and cried at the same time. The blunder's evolution was fascinating, a $1 billion snafu tied to the prestigious optical firm of Perkin-Elmer of Norwalk, Connecticut, which had won the contract to produce the world's finest mirror.

The idea behind Hubble was obvious. Unlike ground telescopes operating below miles of soupy, churning atmosphere, an instrument in the vacuum of space could finally capture the dreamlike beauty of the Orion Nebula and the symphonic tendrils of galactic structure. As if stumbling upon our first optometrist after a millennia-long myopia, we could pinpoint quasars at the edge of the universe and even spy on planets with "being there" clarity.

But the cobalt spirals of distant galaxies were more than merely pretty; Hubble's instrumentation could resolve in them the cosmological issues that had vexed human brains since they first grew large enough to be tormented. It could supply some of the desperately needed hard data that would explode with a loud whoosh into the conceptual vacuum created by theoreticians. It would satisfy everybody.

But could we really pull it off? Could we defy our own bumbling, uneven history and have the raisins without the bran? We are, after all, the species that created exquisite Mayan pyramids and then shoved reluctant virgins from their summits. The ones who carried lifesaving penicillin into one jungle while slash-burning others to produce more hamburgers. Hardly a history suggesting that we could do anything right the first time—especially a project of unique complexity.

And—surprise—we didn't. Hubble's misshapen primary mirror threw a cream pie into the face of the U.S. space agency, NASA, which was blamed for improperly supervising the contractor. In my opinion, they didn't deserve to be the bull's-eye for all the darts; Perkin-Elmer was a prestigious firm, and this was to be their most public, crowning achievement. It seemed inconceivable that they'd blow it. Turns out, when testing the curvature of the optical surface, Perkin-Elmer used an expensive, state-of-the-art device called a reflective null corrector, which told them that the mirror was perfect. But when they double-checked the shape with an older, less sophisticated instrument called a *refractive* null corrector, that analysis revealed a serious misshape. (Serious in this case meant that the outer edges were too low by an amount equal to a fiftieth the thickness of a human hair.)

Now, what would you or I do in such a circumstance? What's the procedure when two tests reveal contradictory results? Check the alignments of both testers? Perform a third test? Here's what Perkin-Elmer did: nothing. They decided that the older instrument "must be wrong." And so a telescope was blasted into orbit with a defect so egregious that most amateur telescope makers would have avoided it.

It cost almost three years, an intricate repair mission, and a half-billion dollars to correct the problem. The irony is that an absolutely flawless backup mirror existed. It had been fashioned by Eastman Kodak and could easily have been substituted before launch, had the spherical aberration been detected in time. And the whole mess could be traced to a single optical element in the fancy reflective null corrector that had been misaligned by the thickness of a dime.

As we all know, the Hubble's corrected optics continue to amaze the world with otherworldy images of such beauty and utility that its initial woes now seem mere labor pains, like those attending any cherished birth. But still, pragmatists may wonder: If such a prestigious, high-profile project can screw up, can't anything?

Can we ever assure future colonists signing up to live on a moon of Jupiter that they won't meet horrible deaths because someone forgot a bolt or a washer?

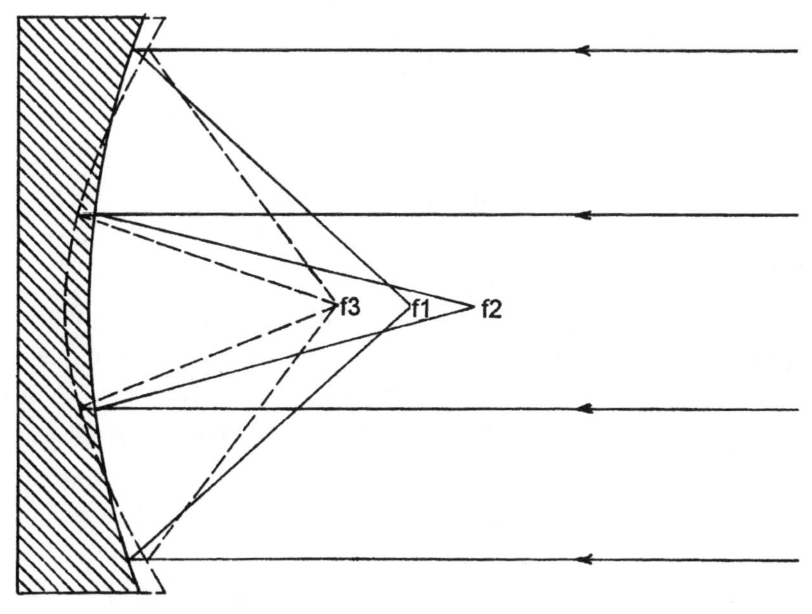

The Hubble Space Telescope's main mirror, shown here in cross section (shaded area), has a defective shape that causes incoming light (arrows) to focus at different places (f1 and f2, for example). A perfect parabolic mirror (Hubble's original specifications, dashed line) would focus all incoming light at one point (f3). This illustration exaggerates the differences for clarity.

No one who heard them can forget the screams of the three *Apollo 1* astronauts when a simple spark, a minor "short," turned their pure-oxygen environment into a blinding inferno that melted their space suits around their bodies.

People continue to speculate about the four-minute-long anguish of the *Challenger* astronauts, that handsome, winsome crew, as they fell, still conscious, into the sea. Or the Russian cosmonaut Vladimir Komarov, whose craft impacted the ground at 400 miles per hour after a malfunction. Reportedly, he wept while speaking to his wife from orbit, knowing how badly things were going and the inevitable outcome that lay minutes away.

These disasters and a myriad of future foul-ups must be part of the equation when we contemplate human space travel. Even the horrors of aeronautical accidents *within* the atmosphere have been made more evident, and personal, over the past twenty years by the presence of CVRs—cockpit voice recorders. These have allowed the final minutes of the lives of captain and first officer to be aired publicly for all to hear.

Previously, the last words of our fallen soldiers have been largely unknown, messages poignantly evaporating into oblivion. Now it's different. Available from the National Transportation Safety Board, CVR transcripts provide not just gripping if morbid reading, but precious insight into the evolution of errors.

The full, slow-motion explication of fatal accidents involving commercial jetliners requires a careful study of the complete transcripts and NTSB expert analysis. Still, consider just these last words, all recorded during the final ten seconds before impact:

April 4, 1977 (Both engines dead after ingesting hail):

Captain: Ah, we're putting it on the highway. We're down to nothing.

First Officer: Oh, Bill, I hope we can do it. I've got it. I've got it. I'm going to land right over that guy.

Captain: There's a car ahead. . . .

The explosion of the space shuttle Challenger *looked so much like other rocket launchings, it was hard to believe we were witnessing a disaster.*

December 28, 1978 (Out of fuel):

Captain: They're [the engines are] all going! We can't make it!

First Officer: We can't make anything. One seventy-three, Mayday! We're . . . the engines are flaming out. We're going down. We're not going to be able to make the airport!

January 13, 1982 (Ice on wings, improper engine setting):
Captain: Forward, forward. Easy. We only want five hundred feet. Come on, forward . . . forward. Just barely climb.
First Officer: Larry, we're going down, Larry . . .
Captain: I know it.

June 7, 1971 (Landing in fog, hit beach cottages):
Captain: All right. Keep a sharp eye out there.
First Officer: Okay. Oh, this is low. You can't see through this stuff.
Captain: I can see the water . . . straight down.

On the other hand, the fiery incineration of the German zeppelin Hindenburg *was a terrifying sight. Oddly, both airships were surrounded mostly by steam rather than smoke—the frenzied conversion of hydrogen and oxygen to ordinary water.*

First Officer: Ah, yeah, I can see the water. We're right over the water. Man, we ain't twenty feet off the water! Hold it!

December 29, 1972 (Inattentive crew; plane glides into the Everglades):
First Officer: [We did] something to the altitude.
Captain: What?
First Officer: We're still at two thousand, right?
Captain: Hey, what's happening here? I . . .

June 23, 1967 (Bad weather):
Tower: Clear to land.
Captain: Yeah, we're all ready. All we got to do is find the airport.

December 20, 1995 (Mountain ahead, but unretracted spoilers prevent full climb rate):
Captain: Pull up, baby.
First Officer: Easy does it, easy does it.
Captain: Up, baby.
First Officer: . . . more, more.
Captain: Up, up, up.

In all cases, the mistakes had already been made. We're merely witnessing the final urgent moments when the participants become aware of the enormity of the situation. Most poignant is the degree of utter helplessness, the puppetlike quality when the decisive point has been passed and the drama can play out in only one direction (although in two of these incidents some of the cockpit crew survived).

Personal blunders result in embarrassment or unintentional comedy much more often than in real danger. Still, my own incompetence has put my life in jeopardy several times; I was very lucky not to have bought the farm.

Once was during a scuba dive. I decided to leave the group swimming with me, 40 feet down, to see what the real depths were like. Intending to return to the others in just a few minutes, before they'd know I was gone, I first lagged cagily behind, then vigorously swam deeper, following the steeply sloping ocean floor.

At 110 feet I stopped, suddenly overcome by intense dizziness. I was blacking out! With a start, I realized what I'd done, the enormity of my predicament. My too-vigorous descent had knocked off the oxygen balance or—something. Now I was losing consciousness. I was far from the others, and my next breath would be water. I was as good as dead.

Panic gripped me. This was it! Done in by stupidity. What should I do with my last seconds of consciousness? Try to relax to attempt acclimating to the depths? Or briskly swim upward—an established no-no? But I did head upward, and eventually caught up with the group, who never knew I was gone. I swam on with them in stunned silence, my dizziness gone, appreciating this abrupt gift of life.

Then, a few minutes later, an amazing thing happened. I took a breath, and suddenly it was like trying to suck on a stone. Nothing! No air! I looked at my gauge: empty. Wild swimming at great depths had exhausted my tanks much sooner than normal activity would have.

So now, for the second time in minutes, I was in great peril. No air, at 40 feet. I swam to the divemaster, made the finger-across-the-throat pantomime gesture for danger, and pointed to my gauge. My lungs were starting to ache. Obviously misunderstanding (for nobody could have run out of air in so short a time), he assumed I was merely concerned about the time remaining, and he nodded and swam away!

Frantically, I had to make a quick choice. Either swim back to him and rip out his mouthpiece for a quick breath from his tank (he'd understand *that*!) or try to make the surface. I headed up, quickly. After just ten feet or so, I found air returning! Later I learned that the decreasing water pressure encountered when ascending would indeed release bits of remaining air. Now I could slow the ascent. And I would live.

Those two unreal rapid-fire experiences on the razor's edge of eternity were enlightening as no textbook analogy could ever be. Usually we live in Western society's congenial grasp. Television, movies, children's school plays, mortgage payments. Neither agony nor ecstasy. Individually, we are pleasantly safe, and our errors in judgment at work or at home may cost us time, money, or reputation but rarely put us in grave danger.

A friend's husband, a construction worker, did perish after falling from an I-beam on which he was sitting. He'd moved too quickly, the wrong way, and another worker heard him utter one quick expletive as he slid off. Normally, however, our homes, instruments, and vehicles are constructed so that a single error is not fatal; usually a *series* of mistakes is needed to do us in. That's why an hour of automobile travel offers just a 1-in-800,000 risk of death.

Our fascination with space exploration and our obsession with high technology must not let us forget the human fallibility factor in the fate of future spacefarers, be they colonists, scientists, or orbiting astronauts. The odds of the space shuttle's exploding are now pegged at 1 in 74. Fly the shuttle three times, as some have, and you're staring at a 1-in-25 chance of dying a spectacular death. The total horror of watching your blunder—or someone else's—unfold with stop-action clarity while your life's final moments inexorably elapse will surely be part of the experience of trekking outward toward the stars.

While those who escape can pass the long, lonely months of interplanetary boredom sharing their own close-call tales of terror, stupidity, and transgression.

And now it's your turn.

Naming the Universe

If the lovely Andromeda Galaxy had been given a less mellifluous name, its awesome, unreachable beauty would of course be undiminished. Pluto and the frozen Uranian moon Ophelia, unknown until the twentieth century, would be just as intriguingly mysterious if they remained unnamed. Nonetheless, our labels for celestial objects flavor our thoughts about them in the same manner as the influences exerted by the names of earthly objects—or people: Bob Dylan, Marilyn Monroe, Gregory Peck, and Clint Eastwood were all well aware that their pedestrian birth names lacked the impact and pizzazz of their adopted ones.

Called away to Asia for three weeks in 1997, I got a surprise when my airplane landed in Bombay, a noisy place I've visited periodically for decades. This time, Bombay wasn't there!

India's largest city had changed its name to Mumbai. And like characters in Stalinist-era photographs, all traces of the old name had been erased. For example, what we call French toast had always been known on the subcontinent as Bombay toast. But no more. Now it's Mumbai toast. If Bombay toast was hard to find on a menu before, the new name probably elevates it to the most esoteric breakfast food in the galaxy.

In the sixties, name-changing was a fashionable way to assert autonomy from one's parents. Cheryl became Chandra, while bikers adopted such less pretentious pseudonyms as Street Dog. In hippie and spiritual circles, young people were calling themselves Meadow or Krishna.

We like stability, but we rarely get it. Cities ought to be more reliable than personal names, and yet in recent years Leningrad and Stalingrad vanished, along with entire countries such as Ceylon and Burma. Only Earth itself seems secure from the renaming craze, since nobody has yet seen fit to suggest a name-the-world contest. Too bad. It would be fun to live on Zorak for a change. We can't physically move from our planet, so the only alternative, for a little variety, is to redefine it altogether.

Such intrigue isn't lost on astronomers. Every few decades the universe's objects get a complete makeover. While a few dozen stars retain their ancient names, which mostly come to us from the old desert-dwelling Arabs and thus are almost unpronounceable to Western ears (when was the last time anybody dropped the name of Libra's brightest star, Zubeneschamali, at a party?), a new system installed a mere two centuries ago tried to lighten things up. The brightest star of any constellation, it was proposed by the German astronomer Johann Bayer, would simply be called Alpha followed by the constellation name. This allowed the always-phonetically-cryptic Betelgeuse to become Alpha Orionis. Twice as many syllables, but at least everyone could say it with less insecurity. The nomenclature stuck, and suddenly the sky was home to stars like Alpha Centauri. A snobbier, scientific, upperclass élan suddenly replaced the often dopey sound of the old star names, as Arneb became Alpha Leporis and Scheat gave way to Beta Pegasi.

Moons got new names, too. As the Voyager spacecraft flew past every planet from Jupiter to Neptune between 1979 and 1989, two dozen new satellites entered our consciousness. At first they were assigned year-and-letter designations, until the world's official naming body, the International Astronomical Union, got around to deciding what to call them permanently. That conservative world body, which has never been known for its humor or inventiveness, cannot be bribed: You can discover a dozen new galaxies, each with a trillion suns, and still rest assured that none will bear your name. The IAU will prefer to grant them stupefying letter-and-number labels simply because nearly all other galaxies bear such designations, and consistency within cosmic categories is the IAU's Koran. It took years for Uranus' ten new satellites officially to become characters from Shakespearean plays, even though it was a foregone choice since the original five moons had been named that way. No matter that the world's citizenry will largely remain unaware that they can now telescopically gaze at the moons Cordelia and Puck.

Farther away still, giant gas clouds once had simple names such as the Orion Nebula—until the eighteenth century when Charles Messier added it to his famous list of comet look-alikes and it became known as M42. Then, a century ago, Johan Dreyer's immediately-accepted New General Catalogue gave everything an "NGC" name, turning the nebula into NGC1976. Nowadays, all three designations are used, in the same more-is-better school of thought that has granted us nine-digit zip codes and eleven-digit phone-dialing ordeals.

Not surprisingly, the renaming craze has plugged into the always popular desire for immortality, giving us a wildly successful Illinois-based company that offers to name stars after *you,* for a modest $45 fee.

Since no astronomer, observatory, or anybody else will ever recognize that new name, the science community is generally unamused and considers the whole thing a scam. But it's never been illegal to call something by another name, or even to charge money for it if you can somehow find a person who's willing to pay. Therefore, what the so-called International Star Registry does (send you a star chart with one too-faint-to-see-with-the-naked-eye star circled, along with a "certificate" that the new star will now be called Jeffrey) is technically on the

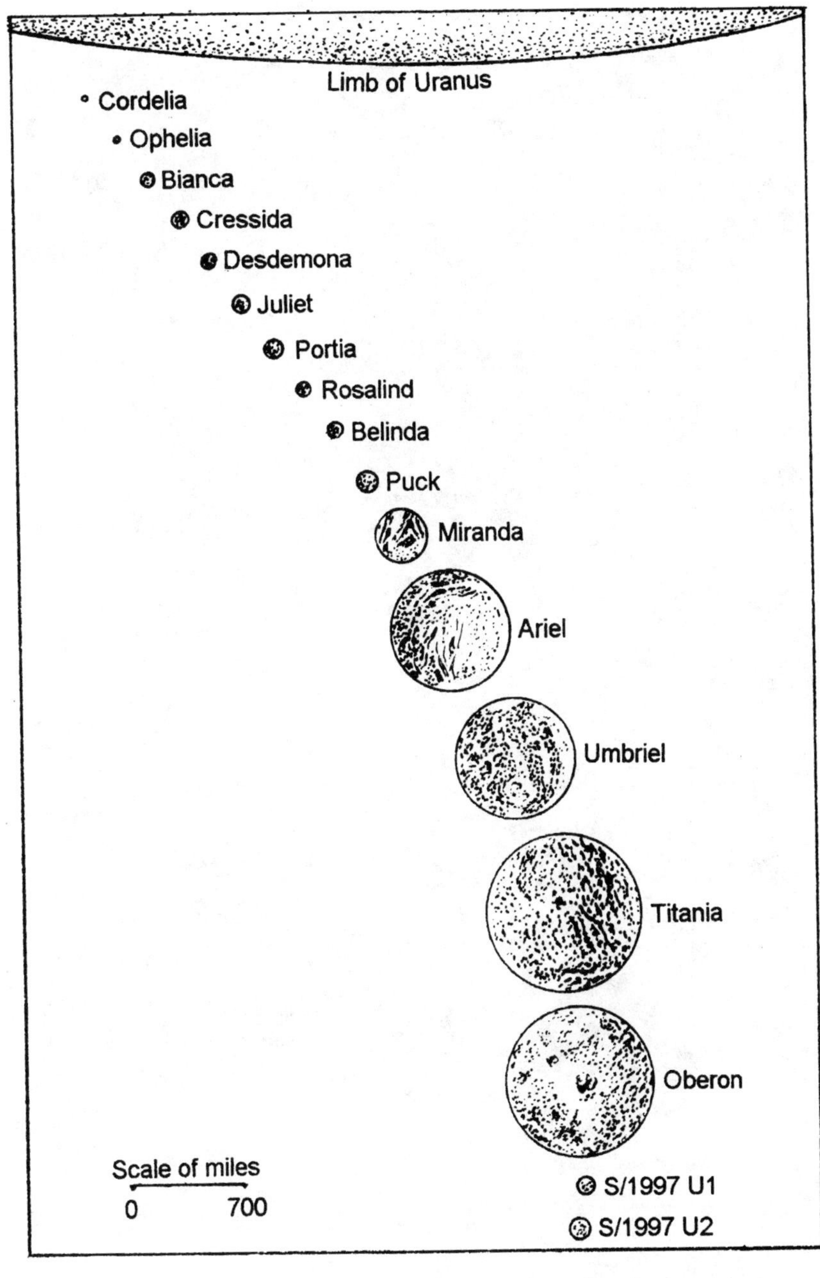

The moons of Uranus in order of their distances from the planet, with sizes to scale. The satellites inside Miranda's orbit were discovered by Voyager 2 *in 1986. The outermost two were found with the 200-inch Mount Palomar telescope in 1997—too recently to have yet been assigned Shakespearean names.*

Twinkle, twinkle, little . . . what? Various names for these celestial objects are concentrically displayed.

up-and-up.* Just as your offer to henceforth call the Brooklyn Bridge by the name Sam Jones is not illegal if Sam actually volunteers to pay you for alluding to that span in that way. It's name-buyer beware.

I'm a little surprised that the Illinois outfit hasn't offered, for a somewhat larger fee, to rename the moon. Or sun. Wouldn't it be flattering if people around the world said, "Don't stay out too long, dear, you'll get a Bettyburn"? In fact, why not the entire universe? Just think, send them a paltry ten thousand bucks, and the universe will henceforth be known as Jeff.

You gotta enjoy the hubris of it. Reaching the age of intellectual curiosity, intelligent youths would ask, "Does Jeff just go on and on, or is there an end to Jeff?" Cosmologists would wonder if, beyond all that is known, there might be other Jeffs.

I've always thought that the bored, wealthy class represents a rich mother lode that ought to be mined. If they're willing to shell out astronomical sums to see foundations or hospital wings bear their names, why allow that one enterprising company to rake in all the celestial bucks? The world's real astronomical body, the IAU, might try being inventive and garner some well-needed dollars by sponsoring a name-the-universe contest. (There's already a bizarre Miss Universe contest, but, okay, the cosmos has got to have room for more than one.)

The real moneymaker, however, would lie in ownership. In an uncharacteristic bit of restraint, the first-to-get-there Americans decided not to claim ownership of the moon. Maybe if the Apollo program had occurred in the days of the British and other empires, we wouldn't have been so circumspect. Back then they had chutzpah: You'd land on a new island or continent, probably through some navigational error, and simply proclaim that it was yours. But not now: We didn't even lay claim to the single desolate acre on which each of the six Apollo landing stages were parked.

You can be sure that some Middle Eastern sheik or other bil-

*Well, maybe not. In 1998, the New York City Department of Consumer Affairs issued a violation notice to the International Star Registry for engaging in a "deceptive trade practice."

lionaire would, out of sheer pride, pay almost anything to own the sun. What a feeling of accomplishment, of achievement, to be able to throw open the curtains as your child awakens and say, "Time to get up my boy. Our sun has risen!" It's perfectly harmless, of course, since nobody will ever be able to *do* anything with the sun.

(Nor will we ever be able to do anything *to* the sun. Some have suggested, as an answer to our garbage and nuclear-waste disposal problems, to simply rocket these undesirables into the sun. After all, dangerous wastes would simply vaporize harmlessly—a perfect solution to an increasingly frustrating global predicament. The problem, other than the extremely high costs of such transportation, is the chance that a launch might go awry, sending all that radioactive debris crashing onto some health-food store.)

If we ever *do* get into the business of buying and selling celestial real estate, we'll have to overcome the economically depressing fact that the overwhelming bulk of the sun and planets is simple hydrogen. Its subsequent glut on world markets would be astronomical.

This is one reason the sun is unlikely ever to have much investment potential. While it weighs an extremely impressive 2,170,000,000,000,000,000,000,000,000 tons, the sun is 90 percent plain-vanilla hydrogen, just like Saturn, Uranus, Neptune, and most of Jupiter. The rest is essentially helium, useful for parties and to help visiting astronauts sound like Munchkins. It's "raw material" at its rawest, and as a final straw it all belongs to celestial bodies that don't even offer such amenities as a solid surface on which to land. The value of the sun and most planets, in today's market, is thus very close to zero. Ownership of a celestial body would be a "vanity" acquisition used primarily to impress people.

Some worry that, as in the game Monopoly, the nature of the capitalist system is to make an ever-smaller group of people ever richer, until someday a single person or entity will literally Own Everything.

Not to worry. No matter how imbalanced the gap between rich and poor becomes, no amount of money or narcissism will allow anyone to ever own all that exists. For it's literally true that you can't have everything. After all, where would you put it?

ALIEN INVASION

Over a quarter of the U.S. population thinks we are being visited by aliens from outer space. According to surveys, it's a belief that has persisted for decades. While the scientific community dismisses such ideas contemptuously (among many reasons cited: Astronomers never seem to see UFOs), invasion by extraterrestrials remains a topic of immense popular interest, especially in places like my hometown of Woodstock, New York.

That's only to be expected, given the unending stream of *E.T.*-like movies, pseudo-documentaries, supermarket-tabloid headlines, and abduction literature. Counterbalancing all of this? Nothing whatsoever. Reasoned discussions delving into the pros and cons of possible alien visitations are simply not part of most school curricula. Nor do the media find it profitable to present professorial scientists patiently explaining why they see the whole thing as misidentification, fiction, or the activities of publicity seekers and the mentally unbalanced. For most, scientific discussion and step-by-step reasoning is boring. Abduction tales are titillating. There's no contest.

Still, a future visit by extraterrestrials is not altogether impossible. While interstellar distances are daunting, a surfeit of intelligent life probably abides within our own galaxy. What might *really* happen in such an encounter?

Our collective expectations have been so thoroughly shaped by the surprisingly consistent fictional versions that such novels and movies furnish an almost mandatory starting point. The place to begin is the obligatory scene where the military commander explains why the aliens from space must be kept top secret: "So people won't panic." As the plot unfolds, however, everyone *does* learn about the invasion, and they duly bear out the general's prescience by running through the streets.

Let's construct a more credible scenario. First, the *New York Times* headline: EVIDENCE OF EXTRATERRESTRIAL SPACECRAFT CITED. But then does our sedentary population immediately stampede through the streets? Crowds of screaming people dashing helter-skelter, trampling others underfoot? Not only does this reaction seem uncharacteristically energetic for a seriously overweight citizenry, but we'd wonder where all these people are headed. To escape alien invaders, no earthly destination would seem any better than any other. Why budge?

In reality, our reactions to an alien visit would be shaped by our perception of their motives. Why have they come? Making war against us would seem unlikely, unless our own aggressive inclinations are the behavioral paradigm for the entire universe—a notion as frightening as the aliens themselves.

Arriving in search of food is just as ludicrous. The more dissimilar the life-forms, the less likely they are to find agreement as to what constitutes tasty cuisine. On a more optimistic note, they might be altruistic and have come to instruct us. But an invasion by didactic pedants might be just as intolerable as an onslaught of creatures looking for a midnight snack.

Perhaps they'd come as salespersons, to sell us things. Since so many of our own initiatives are commercially motivated, maybe other life-forms operate the same way. We're a planet with an unusual amount of lead and uranium. We could trade some of it for new, entertaining gadgets; we'd be the Indians that sold Manhattan island.

But let's back up a bit to see *why* this alien business is so appealing. Is it the human love of mystery and the unknown? Or do the tabloids and TV merely reflect a fascination that derives from the scientific community itself?

Scientists routinely announce lofty, abstract, long-term goals to justify their areas of research. Astronomers at NASA cannot seem to bring themselves to admit that we have sent a machine to orbit Jupiter because it is an amazing place to explore, period. They always add something like, "This will teach us about Earth."

The public is apparently not to be let in on the fact that astronomy is a useless science. There is nothing practical to be gained by learning that Andromeda is really 2.2 million light-years away and not 750,000, a correction gained from the Palomar telescope.

Alien scientist on vacation in the Virgin Islands. An extraterrestrial investigator might focus on rare and threatened species first, ignoring humans, who, as a dominant species, will be around for a while.

The revision, so important and dramatic to astronomers, will not affect even one of the 6 billion people who have the potential to learn that fact, and it won't improve their quality of life in the slightest. That we are taxed so researchers can glean information like this is wonderfully pointless.

Faced with the choice between abstract knowledge and cold cash, relatively few taxpayers would agree to fork over ten bucks apiece to send a spacecraft to the giant outer planets. So we don't ask them. NASA's policy: Don't ask, don't tell. When we talk about space missions, we say it's so we can learn about Earth: Discovering more about our planet puts a good face on the investment.

But the ultimate "amazing discovery" toward which everything else is supposedly vaguely directed, the finding we are told will Change Everything, would be to find that We Are Not Alone. Uncovering extraterrestrial life, it's repeated as gospel, would be the Greatest Thing Ever.

Why? And who says?

When it was discovered earlier this century that some life coexisting with us on Earth may be just as smart as we (for example, dolphins), did that Change Everything? Hardly. We have not even catalogued most species of terrestrial life. We don't know what they do or really understand their role in the world's ecosystem. Why should life beyond our atmosphere be such a quantum leap, more important than life-forms we either take for granted or routinely bring to extinction in the name of cattle-raising and hamburger production?

So, let's say the aliens come. First off, we'll want to know: What do they want and what's in it for us?

We'll assume that they're smart, so right off the bat we'll attempt to exploit their knowledge, and we may well be unsuccessful. We've always been bothered when anyone has inside information about which we're left out in the cold, so uncommunicative aliens with patently superior knowledge or abilities will simply bug the hell out of us. *Annoying* might ultimately be the most accurate description of an invasion by benign but taciturn aliens.

If they're passive and prove as tasty as lobster, then another outcome follows: We'll eventually eat them.

If they're vastly superior militarily, we'll be forced to accede to their whims. Or they could simply keep us in the dark about their motives, their reasons for remaining in our lives as unfathomable as that of macaroni salad.

If they're innocent and naive, well, heaven help them. They're out-of-towners stepping off the bus at Times Square. After we do the old three-card monte, they'll be lucky to go home with the ship on which they came.

But run through the streets? No, that's not our thing unless we're looting and burning. And we don't do that in novel, puzzling situations. When something truly new occurs, our most likely reaction is to reach for the phone to be the first to tell our friends.

The fun part, of course, would be if we could communicate with them using anything resembling language. Many of us have fairly close relationships with favorite mammals such as dogs, cats, or even horses. But no verbal communication. We don't *want* our animal friends to speak, even if they could manage it. Who would want Spot to say, "Hey, Jim, c'mon now. It's six A.M. and I wanna go out and pee on your car's front wheel"? Or a cat to sniff the new food and say, "This stuff sucks, Joanie"? We seem to enjoy the dog's obsequious whines and definitely get a somatic buzz from the cat's purr. In short, we're very happy to enjoy their own vocalizations.

Not so with aliens. In no popular film are we shown taking the time (and it might be centuries) to "feel them out" and determine their likes and dislikes, to merge with their "vibes." Most of us would have little patience or inclination to learn to love alien sounds and habits. Instead, we expect aliens to communicate with us in human terms, even if the first efforts involve strange noises and unsuccessful translation attempts. However it's managed, what we really are seeking is not meeting yet another life-form—for we have a surfeit right here that we will ignore right until the last individuals are extinct—but *verbal* companionship. This is something humans have never had with another species. It is this lonely hearts club quality of earthly life—the need to know we are not galactic orphans—that tacitly drives the space budget, the unacknowledged bottom line in extraterrestrial exploration.

Alas, here too we have not thought the matter through. Just as the new car soon loses its charm and novelty, the new conversations will be valuable only so long as they produce useful information. Can they give us cures to diseases? Faster rockets? More powerful weapons? What's in it for us? Once that issue has been milked, we don't want to sit around chatting about the weather on the planet Vandor. Scientifically oriented types, or those who love exploring the endless vagaries of our own natural world, might want to gather around the campfire to discover their "tales from the beyond." But most people would probably prefer sticking with neighborhood gossip. If we do not presently enjoy hearing about everyday life in Tibet, or learning a little geology or meteorology just for the sheer joy of it, then what, really, will the aliens have to tell us? Can they instill joie de vivre in the chronically bored?

(Ah, but if they could, wouldn't that be the greatest gift they could possibly bring?)

The metaphysically minded would want to hear the spiritual views of the extraterrestrials. Do they believe in God? Do they subscribe to reincarnation? Do their views match those of mystics through the ages who claimed that truth is universal if ineffable?

But here again we can smell trouble. Rabid members of organized religions together with the church hierarchy would feel threatened by the far-out competition. How likely would it be that our three-eyed visitor from Nereid would be a Presbyterian? No, these are heathens whose views must be rejected. It's hard to see how their philosophical counsel would be welcomed.

Extra trouble: What if they offered more than mere counsel?

What if they actively came to proselytize? Celestial Jehovah's Witnesses who had not only found a new neighborhood, but the ultimate in a captive audience—6 billion humans held in place by an unyielding gravitational field.

Many would follow them if they played their presumed ace of advanced science. Technology just modestly beyond ours would seem indistinguishable from magic. If they could levitate or walk through walls, many of us would sit at their feet as if they were Hindu swamis. Just as the Connecticut Yankee in King Arthur's Court ruled the masses with a few wizardlike items such as matches, aliens could wow our socks off with the same kind of gadgetry we ourselves might possess a half-century hence.

Or perhaps their interest would involve real estate. In a universe where most surfaces are either gaseous or too tiny to hold on to an atmosphere, we live on a rare and therefore valuable property. Moreover, Earth is the only known planet with liquid water. Aquatically oriented aliens might find our terrain as irresistible as a water-slide theme park.

Location is everything, and we occupy this solar system's prime lot. Any alien interest could drive up property values. Or, conversely, the global insecurity created by alien attention could cause landowners to sell off and invest in whatever suddenly gained value because the aliens had no interest in taking it.

Of all the possibilities, most sci-fi movies seem focused on scenarios dealing solely with war and hostility. And it is the general paranoia that such fiction has collectively instilled that may flavor our response if the event ever does unfold. Aliens may be anything; their motives may be anything. But we will surely stay in character and greet them with the suspicion accorded strangers nosing around the Ozarks.

It is possible that visiting aliens would be tiny, perhaps even microscopic. It is even possible that we have already been "invaded," aeons ago, and that these aliens are now incorporated into our lives.

Or even our bodies. Our very existence is possible only because energy-producing life-forms called mitochondria live within each of our cells. Mitochondria are so different from nearly all other kinds of earthly life that they are alien in almost every sense. Could they have landed here after hitching rides aboard comets or meteoroids and eventually found their niche not only on our planet but within the same brains that contemplate the strange affair? Such symbiotic relationships may be the ultimate result of meetings between life-forms of different worlds—but only if they are both carbon-based and structured around amino acids.

It is also possible that aliens could be altogether invisible to us. Since 90 percent of the universe is made of unknown material, aliens composed of such substances would have properties of which we could not even dream. But even if they were built of baryonic matter—atoms just like ours—they might experience *time* in a radically different way. Just as we cannot see a hummingbird's wings, a creature that moved through time at, say, a billion times our own rate, would be extremely difficult to detect.

Suppose aliens might have found a way to blend in with any background at all: extraterrestrial chameleons. Then, too, our eyes are sensitive to just a narrow part of the electromagnetic spectrum. Creatures who could somehow radiate solely in the infrared or microwave range would not be seen.

If amino acids do form the basis for life in other places, then our detection of them in meteors and, more recently, in amino acid precursors in distant nebulae augurs well for some kind of universal biological commonality. On the other hand, the differing atmospheres, gravitational fields, compositions, temperatures, biological defense needs, and degrees of competition with their own native biota would surely dictate form and function resembling nothing of earthly life at all.

The bottom line is this: There is no way, extrapolating intelligently from our terrestrial situation, to gain the slightest idea of what alien life would be like. Having been wrong repeatedly about much simpler things (such as how the far side of the moon would look), we can only plumb our own tendencies to see how we our-

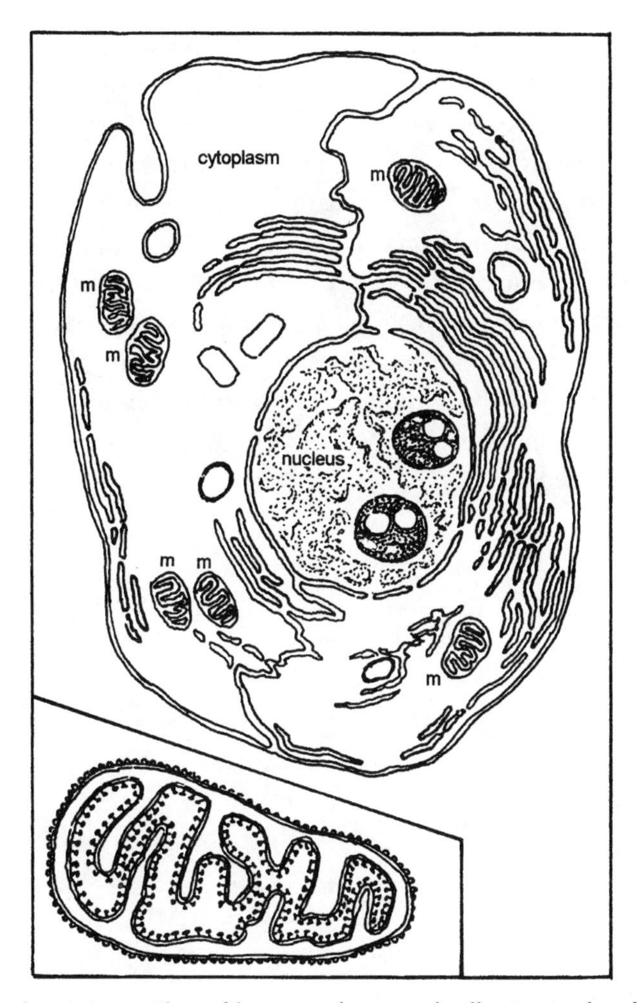

Mitochondria (m) provide usable energy for animal cells. A mitochondrion is shown enlarged in cross section (lower left) *to reveal the minute particles on the outer and inner surfaces that oxydize glucose and store the resulting energy as adenosine triphosphate (ATP) molecules for later release. How did these creatures find their way into our bodies?*

selves would react to the visit. And perhaps live more in the present moment by honestly acknowledging that finding other life is *not* the ultimate goal of space science.

The real goal is exploring for its own sake, for the sheer joy of it, and letting new findings take us where they may. This is a nonstructured, open-ended procedure that seems capable of stirring our minds and spirits indefinitely, a process that works simply because it has its own life. Discovery's winds carry us in its maniacal eddies, propelling us along very different paths than we'd be following were we pursuing any static, charted purpose—including the Finding That Will Change Everything.

Meanwhile, my idea is to set up an interplanetary buoy, like a barking watchdog: CAUTION—QUIRKY, DANGEROUS ORGANISMS ON SOL THREE. Then our consciences would be clear. If aliens come anyway, and don't have the good luck to land in a UFO-receptive place such as my beloved Woodstock, well, they can't say they weren't warned.

It's a Gas

So much of our modern world is made of synthetic materials, we often lose our understanding about the nature of things. Not too long ago, the objects we'd grasp were metallic, ceramic, wood, or stone. That was before we were overwhelmed by thousands of different plastics, fiberglass, composite boards, and the like. No wonder we've largely given up trying to figure out the makeup of the stage sets in which we conduct our lives.

But what about the true basics? As we seek to understand the cosmos, can we profoundly ignore the nature of earthly materials, such as the air we breathe and the gases that envelop and permeate us? A mere century ago, some of our most brilliant minds embarked on precisely that quest of discovering the nature of our invisible atmosphere, that preamble to extraterrestrial exploring—and found it surprisingly frustrating. Equally amazing is how their discoveries have remained unknown to the vast majority. Come along, let's step on the gas for a vaporous adventure.

From birth to death we inhale mostly nitrogen. It composes 79 percent of the air. It doesn't hurt us and it doesn't help us—something like those Hare Krishna people at airports.

Air's second component (one fifth of it) is oxygen, high on everyone's "favorite element" list. It's the *only* element people can buy ready-dispensed from machines, in places like Tokyo. Though only the third most prevalent element in the universe, oxygen managed to amass itself like concentrated orange juice—in greatly condensed form—here on Earth, where it has wormed its way into most surface rocks and makes up two thirds of each bit of sand and quartz.

So much for 99 percent of the atmosphere. What's the other 1 percent of the air we breathe? Put another way, after nitrogen and oxygen, what's the most common element you have inhaled, nonstop, since the moment you were born?

Guess.

"There are two things, science and opinion; the former begets knowledge, the latter ignorance," said Hippocrates in 440 B.C. A few thousand years later, what most of us don't know illustrates the poverty of contemporary scholarship: The average citizen is unaware of even such basics as the distance to the moon and the components of air.

Most would guess carbon dioxide, but that's not an element, and anyway, it accounts for less than one tenth of a percent, despite heavy greenhouse-effect notoriety. We assume carbon dioxide is a major atmospheric player because we're inclined to pay attention to villains, and CO_2 gets unrelentingly bad press. Understandable. Carbon dioxide, which we exhale, is the gaseous equivalent of urine. We don't want it, and the fact that plants like it only diminishes its status. After all, they enjoy ammonia and horse manure, too. Plants are the dummies of the planet. If they like something, we shouldn't like it. So, no, carbon dioxide is not the third component of our air, and neither, except in very humid air, is water vapor. Anyway, the latter varies so greatly from place to place and hour to hour, unlike everything else in the atmospheric witches' brew, that we must ignore water as a reliable component.

Hydrogen? No way. Although there's four times more hydrogen throughout the universe than everything else combined, you wouldn't know it from what's doing here on Earth. Unlike our oddball planet, places more compositionally characteristic of the cosmos (spheres like Saturn, the sun, or the vast, beautiful nebulae) are overwhelmingly hydrogen, and after that you'll find helium. But not here.

Whatever bits of leftover terrestrial hydrogen didn't protectively find a secure niche by linking up with other things (as, for example, in the H_2O of water or the CH_4 of methane) found themselves in a precarious perch high atop the atmosphere. There, random motions caused by sunlight bouncing them around like tennis balls leaked them into space, until by now the drip-drip-drip like a faucet with a cracked washer has reduced the remaining free hydrogen to just minor-bystander status in Earth's envelope.

Know the correct answer? Not many do. It's an element discovered only a century ago.

Argon.

Argon is the gas used to fill lightbulbs, so we look at it a lot but think nothing of it. Which is only fair, because argon does nothing to encourage our thinking process, unlike hydrogen and oxygen, the stuff of which nature mostly fashioned our brains.

Argon was discovered by a Scot, William Ramsay, who eventually won the Nobel Prize for his work with gases. He also discovered neon, krypton, and xenon. He even sent electricity through inert gases to produce the brilliantly hued tube lights that now fill our nocturnal hours with nouns such as BEER or short incandescent suggestions such as EAT. Ramsay invented the neon light.

What a simple idea: High voltage would goose the gas trapped in a tube until each atom's electrons leaped into a higher orbit, which atoms don't enjoy doing at all. So at the first opportunity (in a fraction of a second), they'd tumble back down again, emitting a "particle" of light of a precise color.

That was just a hundred years ago. Who's heard of Ramsay today? You'd think someone who'd discovered more of the universe's ninety-two natural elements than anybody in history and was single-handedly responsible for our culture's neon surrealism would be remembered at least during the century in which he lived. (Ramsay died in 1916.)

What would he conclude today, if he could survey the few concise words fashioned of his light? He might wonder why its appli-

Lightbulbs need to be filled with a gas that won't react with the white-hot filament. Argon is perfect: It is abundant, cheap, and inert—which means that the atom, with eighteen electrons, has all the electrons it can hold.

cations came to be so limited despite its widespread use. Why don't police stations have flashing lights saying TURN YOURSELF IN NOW!? Why don't we find humorous signs, like pharmacies boasting WE DISPENSE WITH ACCURACY!? Why do the neon boulevards of today's cities contain so little humor?

There's also nothing funny about fluorescent lights, which glow by essentially the same process, but with radiant phosphors stuck to the inside of their tubes. The fact that no natural light—not starlight, moonlight, or sunlight, and neither meteors, rainbows, nor lightning—emits light by that process explains why neon and fluorescent lights feel so alien. They *are* alien. Ordinary incandescent bulbs do mimic the process by which nature emits some of its light. Fluorescent does not. Any wonder a day at school, the office, or the mall feels like a visit to the planet Winderznift?

Also, just for the record, not all "neon" lights contain that gas—only the orange and red ones do. Most of the other colors in so-called neon tubes are produced by colored glass surrounding—you guessed it—argon.

But argon doesn't have much of a place when it comes to the solid mass of Earth. It's not a heavy hitter in any other planet's atmosphere either. The four giant planets boast boring brews of mostly hydrogen compounds, without even the dreary compensation of inert gases for a little variety. Our small sister planets each offer little else but carbon dioxide—crushingly hot and thick on Venus, cold and thin on Mars. It takes 100-mile-per-hour Martian gales to rouse enough oomph out of that skimpy air to raise planet-wide dust storms that periodically reduce that world to an impenetrable mist, erasing all surface features as if our telescope mirrors were suddenly steamed over.

So Martian air reveals its presence, just as all but the single atmosphere-free planet, Mercury, betray the existence of a gaseous shroud, armor against the hard vacuum that extends all the way to the edges of space-time.

But not here. Unlike all the other planets, Earth has an atmosphere with transparent ingredients. True, water vapor often con-

A 50-mile-per-hour wind in the skimpy atmosphere of Mars is unable to lift the astronaut's flag. How much wind, then, did it take to form the dunes in the canyon floor?

A 100-mile-per-hour wind stirs the flag but impedes the astronaut no more than would a 9-mile-per-hour breeze on Earth. Martian dust storms are associated with 300-mile-per-hour winds, according to evidence from Mariner 9.

denses out to form clouds, but clouds are not gaseous, they are droplets of liquid. Water vapor is completely transparent! (The steam from a tea kettle is invisible for the first few inches from the spout. This is the actual steam. Only after it starts condensing into a misty liquid does it get whitish, which leads nearly everyone to assume that steam is white.)

This sometimes-opaque quality of H_2O, as well as its "feel" against one's face on a foggy day, makes people think that steam, fog, or mist is heavier than the rest of the air and makes the floating clouds seem so illogical. But contrary to "common sense," water vapor is thinner than every other major component of our atmosphere. Dry air is actually much denser than moist air. That's why airplanes require faster speeds and longer runways to take off in foggy weather: Humid air is *thinner,* and offers less substance for wings to push against.

The transparency of our atmosphere helps explains argon's ability to hide from the human mind until a mere century ago, despite permeating our lungs since our original ancestors first inhaled.

Meanwhile, the person responsible for our thoughts about it has vanished from cultural consciousness, his fame no more substantial than the gas within his neon tubes. Let's salute him now, imagining his name blinking on and off in bright orange like a beer sign: RAMSAY. RAMSAY.

Unwelcome Surprises

Talk about a love/hate relationship. Officially we abhor disaster; we're appalled that nature can burst dams and bring down passenger jets. Yet nothing sells more newspapers or draws higher TV ratings.

That maniacal, evil smile we must somehow keep from our face when telling about a tragic event is testimony to our secret love for the fury of nature: We cherish the impotence of humans, who value their lives above all else yet are powerless to secure them against Earth's delirium. This madness seems so innate, it's worth a look.

While in most respects our planet is an oasis in the night's turbulent desert, our susceptibility to violence and calamity is par for the cosmic course. There are worlds similar to the moon in which, like congressional subcommittees, nothing exciting has happened for the last 3 billion years. There are also places like Jupiter where violence is routine. (Awesome winds, lightning, and lethal radiation are just a few of its perils.) We're rather in the middle, as indeed we must be. If we were inert enough to be perfectly safe, we'd be in too insipid an environment for chaotic mixtures, and DNA would never have woven its spirochete lanyards.

Conversely, a bubbling, hellish porridge of unrelenting upheaval would quickly break apart complex molecules. Again, no life. So we were destined to live in a fairly atypical, stable place that nonetheless springs nasty and inconsistent surprises upon us with diabolical irregularity. How, then, can we complain? It's the way things had to be.

Most of my close friends juggle income in order to maintain their modest lifestyles, but one particularly affluent couple has a beach house in fashionable Westhampton on New York's Long Island, in which we happily enjoy guest privileges. Every two years or so, a hurricane removes most of "their" beach (which amazingly gets reinstated at taxpayer expense) and my friends lament the inconvenience of having their beautiful summer perch sited on a volatile planet. Their viewpoint, gained over the scant decades of a human life span, makes it hard for them to realize that the narrow spit of sand upon which their house was built is, to the ocean, no more substantial than floating kelp. Unlimited government intervention may preserve that trendy region for the remainder of their lives, but nothing can stop the eventual melting of South Polar ice and the rise in sea level that will convert the neighborhood into aquatic brine. Later, even the oceans will vanish, plunging property values further.

Those same nor'easters that periodically slam into my friends' beach, bearing gifts of howling winds and startling waves, mesmerize any brave onlooker. We who have enjoyed wading the tranquil ocean and listening to the summer surf can't help but be almost hypnotically enthralled by the lunatic fury of a marine storm. Its power is unique, the metaphor against which other frenzies are compared. A single hurricane promiscuously expends more energy in a day than the power contained in all the world's nuclear arsenals. If the price for this explosive meteorological arrhythmia is a few dozen abused houses, we witness it happily (so long as we don't pay the cost).

So let's admit we love disasters. They're a movie producer's fail-safe project. The only remaining question: What's your favorite form of mayhem? Of course, the harm to human life prevents us from truly enjoying natural calamities. Yet, while the heart deplores such pain or loss, the mind accepts it, knowing that disasters are bad: *Disaster* comes from the old astrological belief in unlucky stars—*dis* ("bad") and *aster* ("star"). There's no denying that the feral component of our psyche gets off on such unshackled natural violence. Most of us want to see the wind blow more strongly, the lightning intensify, the waves grow more furious.

As an astronomer, I ought to be drawn to that most violent of

The rapture of observing a wild marine storm is somewhat diminished by being in the middle of one.

all catastrophes, the supernova. Imagine the scene where a massive star blows itself up with the temporary luminance of a million suns, taking its entire neighborhood with it. Even the treacherous buildup, where matter from a placid companion star slowly settles on the surface of a massive sun, pulled in by its gravity until that perilous layer of explosive hydrogen ignites in a star-wide conflagration—even this prologue carries an alien, menacing, unearthly flavor. It's too much, a violence utterly beyond any human to fully grasp. So while I did journey to Panama in 1987 to get near enough to the equator to see the first naked-eye supernova since the year 1604—just to stare dumbly at a frenzied star that suddenly appeared clear across the sea of emptiness in the next galaxy, 170,000 light-years away—my favorite cataclysm happens right here at home. I cast my sadistic vote with the legions whose most dramatic and therefore favorite catastrophe is the earthquake. It may be simpleminded, but it's the only natural disaster whose upheavals involve solids rather than liquids or gases. Pandemonium involving the latter just doesn't seem quite as concrete.

First, you've got the ground shaking. The ground! The very metaphor of stability, the rocks, the ground of being—rattling around like an unbalanced clothes dryer. Then there's the remote possibility that the earth will part and you'll fall in. Granted, this is a peril sufficiently minuscule to escape the fine print in every insurance policy, but still, it could happen—resurrecting images of Dante's inferno and the classical location of hell. Imagine being captured and sucked in by Earth itself! Wonderful!

But before it begins, we get not violence but intrigue, in the form of *earthquake lights* (although these remain controversial and largely anecdotal). On the one hand, many eyewitnesses have reported bizarre flashes or glows in the sky immediately preceding major quakes, but sometimes a day or two before. Few have been documented scientifically, and given the unreliability of other skylight reports, such as UFOs, we're wise to be skeptical. On the other hand, there is a plausible explanation: The piezoelectric effect, in which masses of subsurface quartz become compressed by

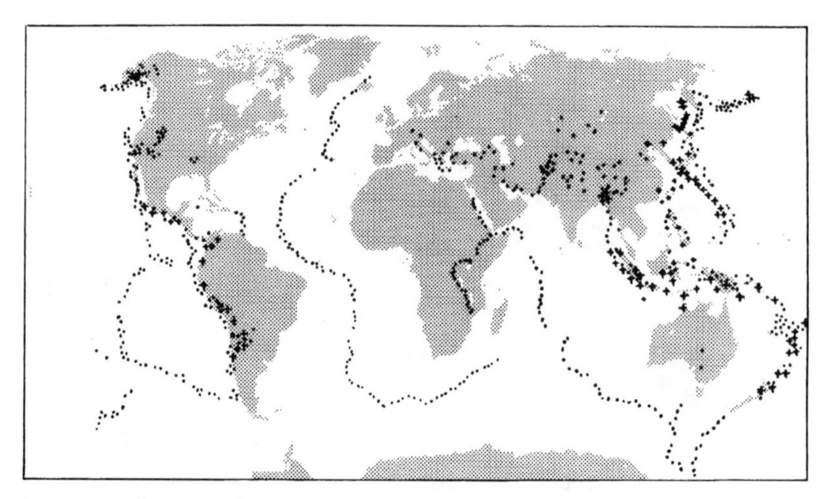

How to find (or avoid) earthquakes, the author's favorite cataclysms. Dots indicate shallow quakes, less than 62 miles deep, associated with tectonic spreading zones and hot spots. Crosses denote deeper earthquakes associated with the subduction of one plate beneath another.

underground motion, releasing electricity. How this can find its expression in the sky is a mystery, but at least we're halfway home in finding a mechanism by which nature can weave a light show linking earth and firmament.

Picture the coming catastrophe with or without a ghostly heavens-spanning light, and definitely factor in dogs, cats, frogs, and other animals sensing strangeness and acting accordingly. Then imagine the mad upheaval of the earth itself. Where can social convention or government decree find a place in this dynamic context? When all the presidents, all the dictators, all the police forces of the world are powerless to delay the scrawniest earth tremor by a single second, we stand exposed and naked in the cosmic wind.

As developed nations acquire greater resources (mainly by taxing us to the molecular level), governments admirably try their best to stay the hand of catastrophe. Dams, breakwaters, levees, and weather satellites successfully moderate nature's intemperance, while catastrophes involving human error slowly yield to ever-longer lists of safety procedures and regulations. Let's take air safety as an example; as a pilot, I'm well acquainted with it and stand forever enthralled and a bit apprehensive at the threads on which it dangles.

Most flying in the United States occurs under regulations called VFR—which stands for Visual Flight Rules. In most airspace, which is "uncontrolled," the key concept is "see and be seen," which places responsibility where it belongs—on the individual pilot.

The "be seen" part of the rule lies out of every pilot's hands, since you have to trust the eyes and attentiveness of others. I made it even harder for them, handicapping other pilots when, in a stroke of what may be terminal stupidity, I had my plane painted sky blue. Gorgeous, but I somehow overlooked the possibility that this would make it blend in with the background.

Even without such a disadvantage, the "see" part of the edict is tricky. Another plane approaching you at 300 miles per hour will become visible only some twenty seconds before impact. If you happen to be looking at a chart or instruments, or turning to a passenger during that critical time, well . . .

Yet midair collisions, which have the potential to ruin one's entire day, rarely happen: only thirty a year out of 38 million flight hours. Chalk that up to the fact that aircraft are tiny compared with the wide open spaces of their environment. But what about the larger bodies of planets, in the even larger zone of the solar system? Now, *those* are collisions that can make us take notice.

On some scale we are continually witnessing worlds in collision. Every ten minutes, on average, on any clear night, a meteor punctures the heavens over every spot on the planet. A house in North America is struck every year and a half. And a big, noisy intruder comes in once a century or so.

It wasn't a disaster, but it was certainly an attention-getter when, in November 1982, a meteor came through the roof of Bob and Wanda Donohue's home in Wethersfield, Connecticut. Puncturing their ceiling with a terrifying roar, it slammed into the room adjoining the one in which they were watching the television show

*M*A*S*H*. The grapefruit-sized party crasher created history: It was the only instance of a town being struck *consecutively* by meteors. (In 1971, a home barely more than a mile from the Donohues' had been similarly impacted.) Why this community near Hartford had been selected for this unprecedented distinction remains a mystery, as well as a model of irony—Hartford is the headquarters of many insurance companies, which, no doubt, would cite the statistical impossibility of such a repetitive occurrence.

The Siberian meteor of June 30, 1908, was probably a thousand times larger but exploded (some meteoroids do that) above a region with lower property values. Its damage was unseen until explorers reached the area years later to find flattened trees for 50 miles in all directions.

Once a millennium or so, an even bigger one comes in, prompting serious philosophical questions about the general stability of the solar system. While the Donohues' insurance policy did indeed cover the damage from the shooting star, an event like the one that destroyed half the life-forms of Earth 65 million years ago may elude such prompt service by adjusters.

A lesser known but far more devastating disaster happened 250 million years ago, in the Permian age. That time it was more than a mere shake-up for our planet—it was nearly a knockout blow. More than 99 percent of all earthly life-forms vanished in an immense collision with a meteor or asteroid. Plants and microscopic animals that had survived for a million centuries suddenly disappeared forever. There's no safe haven in the heavens.

But don't worry needlessly about celestial bullets: The average "shooting star" is only the size of an apple seed, despite its bright pathway in the night sky. The real troublemakers, one to five miles across, rarely clobber us. And that is simply because there's so much room around us—which is why we call it space.

Suppose that we get lucky and manage to miss the next five asteroids of Permian dimensions (each of which could leave the biosphere struggling to recover and erase our earthly scene like dust in a car wash). Even if we evaded the next truly big ones, defying all laws of chance (or else built the anti-incoming-meteoroid defense system now being discussed), our world would survive more or less intact for no more than another billion years. Long range, we cannot win, no matter what.

That limit is nonnegotiable because the sun is steadily growing brighter. In just 1.1 billion years, it's now calculated, it will have become 10 percent hotter. Ten percent warmer may sound like something we could get used to, sort of like moving from Boston to Atlanta. Unfortunately, that seemingly modest alteration will prove more than enough to evaporate the oceans and create a worldwide thick overcast that will trap heat in a runaway greenhouse effect, and it won't allow things to stabilize until thermometers everywhere register a uniform 700 degrees Fahrenheit. That's hotter than an oven, too toasty for even high-BTU air conditioners to handle. End of the run. Show's over. Final curtain. Blackout . . .

A shame. We now know that life began here some 4 billion years ago, and it would be nice if we were merely in middle age, with another 4 billion left. But no: We're 80 percent through our allotted custodianship of this planet. We—earthly life in all its varied and wondrous experiments—are like a sixty-year-old, a galactic senior citizen. True, if we identify ourselves as *Homo sapiens,* then, Holy Geritol, we've still got lots of time ahead. But if we see ourselves as life itself, then we're well past our prime.

If humans survive to that age when all earthly life must perish, will we display the sanguine resignation typical of old folk who have seen it all and are ready to move on? Or, perhaps, will we find innovative ways to pack up our terrestrial belongings and relocate to the nearest hospitable world, which is the now-standard sci-fi vision for our future? But why even ask? More than this language itself will be unrecognizable in just a few hundred millennia; our very ways of thinking, our three-kilogram brains and the elaborate circuitry through which our neuroelectric currents connect, will be dynamically changed. No question asked today will generate an answer fathomable in a few million years—long before contract-canceling disasters are likely to happen.

And if uneasiness about the death of Earth a mere billion years from now seems unduly anxietal, what about the current fascination with the fate of *everything*? Why do we bother with cosmological inquiries such as whether the universe will keep expanding?

Perhaps it can't be helped; maybe it's a question asked on every inhabited planet in each of the 50 billion known galaxies, a universal theme song that starts playing the moment creatures realize that a Big Bang started it all. Or maybe it's just a human eccentricity to spin our wheels considering issues that cannot possibly affect us or our descendants.

In terms of new-star formation or the lack thereof, it will be at least 5 billion years (and more probably 10 or 15) before it makes the slightest difference whether the universe stops expanding or keeps right on going.

Bottom line: If ever there was an epic drama that simply doesn't matter, this is it. The universe's death is a disaster as irrelevant as if a university of dolphins debated whether the White House lawn should be mowed Tuesdays instead of Thursdays.

We may be intrigued simply because we're so accustomed to staking out a preference; we demand to know our options at disaster time, even if it's a billion years hence. Which do we root for? That ultimate cataclysm of a closed universe where everything that exists crushes itself down to the size of a basketball? Or its alternative—a cosmos of emptiness, universal death, and absolute cold? It's like choosing between eventual cremation or burial: Neither really sounds like how we want to spend the afternoon. Why then should a similar but universe-spanning Hobson's choice seem so compelling that it recurs like a musical coda in the mass media?

(Incidentally, while *Hobson's choice* has come to mean "no choice at all," its origins date to a seventeenth-century liveryman named Thomas Hobson, who insisted that customers take the next available horse regardless of quality. In short, originally it meant "take it or leave it." Just thought you should know.)

Not only do we have no choice, but "we" will not even look at the eventual outcome the same way when we get closer to it. Current cosmological questions are admirably grand but will surely prove as irrelevant as when the ancients hotly debated the whereabouts of the oceanic location where the world supposedly dropped off.

Our human fascination with disaster runs so deep, we're keen on savoring future ones now, in our imaginations, simply because we know we won't be there to enjoy them.

We can only hope that our descendants will not have evolved so much that they'll lose our childlike, possibly unique-in-all-the-universe infatuation with havoc. For it is really true that coexisting with our heartfelt need for security is a wild, mischievous, lunatic urge to knock down the sand castle, to scream into the galactic storm, "Go ahead! Hit me with your best shot!"

Given enough time, the universe will oblige.

Everyday Alchemy

Like the intricate designs a creative child can fashion from a simple set of Lincoln Logs, the universe has an alchemic ability to concoct wildly different things from the same material. At first it seems disappointingly limited that the same ninety-two elements are the whole story, the entire repertoire of substances no matter where we look in space and time. Only when we pause to observe the startling diversity that pours like a Niagara from any of these elements are we reassured and even enraptured.

It happened to me in 1996 in a dramatic way, while I was exercising my passion of flying, that uniquely twentieth-century hobby. One winter night, I was in my beloved single-engine Cherokee 180 en route to do a lecture with John Dobson, a whimsical astronomer whose wilder ideas we will come to shortly. My flight path took the noisy four-seater past New York City. I didn't know it yet, but the trip was to be a vivid demonstration of this great talent of nature to develop totally different creations out of the same material.

Above the Statue of Liberty and skimming alongside skyscrapers at an altitude of just 800 feet, I gawked at the shapes and lights the way a Martian farmboy might, before continuing up the ice-covered Hudson River. Thirty minutes later the terrain changed to snowy squares of frozen blue moonlit fields that provided a vivid contrast with the red cockpit lights. If there is such a thing as magic, this was it.

An aurora blazed to the north in front of me, as they do with surprising frequency in upstate New York. Constellations shone vividly, while below sprawled even brighter islands of lights marking the series of prisons that dot the rural landscape. Seen from great distances, they blaze with false cheerfulness, navigation beacons for airborne creatures—pilots and perhaps even birds—who travel northward from the sea. Their familiar names—Sing Sing, Napanoch, Dannemora—sound more lovely than those of the stars, whose mostly Arabic origins ring harshly in Western ears. Mellifluous names for ugly places, I thought, a reminder that there are two hundred times more inmates in our country than all the stars visible to the naked eye.

As I continued, the fog that had covered Brooklyn near the ocean gave way to clear air, with just a few high clouds above. At the same time, water in other forms, ponds and lakes of ice, was passing beneath the wings. Except for a strip cleared by icebreakers, the Hudson River was solid as well. This was reassuring: Every frozen body of water is a potential landing site in a pinch.

But although it looks inviting down below, water is every pilot's enemy. We shudder at the thought of thunderstorms, icing on the wings, water in the fuel lines—we want no part of water even as the very brains formulating that aversion are themselves mostly composed of it.

And yet, in other forms, water is a valued companion. As cloud, water droplets give shape to the sky's otherwise unrelenting emptiness. Climbing through giant breaks to soar atop mutating cauliflowers roiling slowly in the bright blue moonlight accords the aviator wordless joy, even when seen for the thousandth time.

So there I was that brilliant night, enveloped in my isolated world, far from the reassuring realms of human intervention, gliding like some interstellar drifter amid wildly different materials all fashioned from a single substance.

That material also happens to compose three quarters of the universe. Plain hydrogen. What could be less interesting? Hydrogen, normally incapable of holding our attention for more than a few mandatory moments in chemistry lab, has managed to make us wide-eyed only a few times in our lives. When the *Hindenburg* and later the *Challenger* exploded, the fascinating and horrible spectacles were actually demonstrations of the simplest possible chemistry. Here was hydrogen releasing itself from a human cage to find its way to its favorite companion, oxygen. Their eternal offspring is water, so that during the explosions, the white, billowing "smoke" surrounding both dying airships was nothing more than—cloud.

Hydrogen caught our attention again in the early fifties with the invention of the hydrogen bomb. Once more, the frightening yet beautiful mushroom cloud was merely the result of hydrogen's alchemic ability to form something else—in this case, helium.

Sunlight is a direct manifestation of hydrogen's creativity, and so are the rings of Saturn and the red nebulae so intricately airbrushed throughout our galaxy's spiral arms. We rarely think of hydrogen, and yet it is everywhere, disguised as if at a costume ball by its ability to manifest itself in so many different ways.

Eventually a double line of runway lights drew my attention, a blazing "11" that, for aviators since the early biplanes, produces a Pavlovian smile. With all the modern instrumentation and navigation equipment, the end of each flight still involves the small detail of finding the airport. Often, its lights are obscured by trees or overwhelmed by the neon blaze of a surrounding city. Here the runway's illuminated "11" was the welcoming buoy of my home base, a rural airstrip named Skypark. After five P.M. it's unattended: no guards, no fences. The main worry is not an intruding vandal or thief but the threat of running into deer on the runway or skidding like a tobogganist on patches of ice. I applied the brakes gingerly; the perils of water are never very far.

I clambered out of the plane and (having cheated death yet again) looked around at the area's natural beauty. In a region with three feet of annual rainfall, nature's canvases are all dominated by watery brushstrokes.

A curling wisp of cloud drifted above, its arabesque patterns mutating like the letters of an alien language. Below, a line of blue icicles hung menacingly from a rock ledge. Nearby, a whirlpool in the rapids sent its white water roiling in a thousand liquid designs.

Hydrogen's magnum opus, then, is its role as the major component of water—its finest performance so far as Earth and humans are concerned. In the wheels-within-wheels shadow play of the cosmos, chameleon hydrogen becomes a compound that itself assumes a thousand forms. Because it envelops us so completely, the single substance called water is worth more than a passing thought if we are to understand this key underlying quality of changeability that so dominates the cosmos—and constantly reshapes human destiny as its mutations spill over into our personal lives and everyday events.

Receiving philosophical counsel from water comes easily to us; our planet seems chosen by nature to exemplify its watery universe. But not because Earth holds some unique status or special dispensation; the most common compound in the entire cosmos *is* water.

This is no surprise. Water is fashioned out of oxygen and hydrogen; the latter is the universe's most abundant element, and oxygen, though a thousand times less prevalent, is the substance that overwhelmingly loves to mate with other things. Small wonder their courtship and perennial "exchange of rings" should be repeated in every corner of space and time.

Gazing beyond our atmosphere, we see water abounds there, too. Comets are largely balls of ice. The moons of many planets are globes of solid ice, nearly to their cores. Most of the night's stars stand enveloped in gaseous water vapor: steam! Water in all its forms recurs like advertising circulars sent to every isolated galactic hamlet.

Commonplace, yes, but water is also very peculiar. When changing from a liquid to a solid, almost everything in the universe contracts, growing denser in the process. Water bucks this rule by expanding and becoming lighter—an oddity that has proved a major nuisance. If water behaved "normally," our lives would be very different. Icebergs wouldn't float (the *Titanic* would not have sunk), and nobody could ice skate except where a body of water had completely frozen from surface to bottom. Pipes wouldn't burst, either.

Even more peculiar is that we live on a planet whose average temperature hovers just above that critical 32-degree-Fahrenheit mark much of the time. After all, ice exists for a generous 500-degree range (from absolute zero at −459.67 degrees Fahrenheit to +32 degrees Fahrenheit) and steam manifests for thousands of degrees, all the way to where motion becomes so jittery that its molecular bonds break apart. But liquid water inhabits an extremely narrow range of just 180 degrees, which happens to embrace our environment and our bodies. It's only to be expected, then, that liquid is the state with which we are most comfortable and familiar, while ice and steam are the real templates for water's endless creative designs almost everywhere else in the universe.

But a waterworld needs more than water plus the right temperatures. It can exist only if it boasts a substantial atmosphere as well. Water, the bully that wears down mountains and tosses ocean liners, becomes a feeble invalid when air pressure is reduced. This effect has disheartened many mountain climbers who have watched their Coleman stoves appear supercharged as they made melted snow boil long before reaching 212 degrees. Consequently, no café is likely to succeed atop Mount Everest; it's impossible to make a hot cup of coffee above 20,000 feet without a pressure device.

The restaurant potential decreases further on worlds with skimpier atmospheres. On Mars, water evaporates so willingly that it will always automatically boil into vapor and vanish; that is, if it doesn't freeze first. You can't have liquid water there even if you import it! This is why the many meandering fluvial channels discovered on the Red Planet by the Viking spacecraft in 1976 were so astonishing: They tell us not only that Mars once had copious water but that its atmosphere was much thicker than it is now. Might the small rocky world go through cycles where it takes turns being alternately friendly and hostile to life? Could creatures from long ago have burrowed beneath the Martian soil, where ice (and possibly even liquid water) still remains?

But none of these oddities or otherworldly possibilities can hold a candle to water's strangest, simplest, and least appreciated characteristic. It is this: The two hydrogen atoms chemically bonded to one oxygen atom that constitute water are linked not in a straight line (180 degrees) but at an angle of 105 degrees.

This fact alone has made life on Earth possible (and perhaps on endless other worlds as well). The angle of 105 degrees allows the water molecule to boast a polarity where the oxygen portion has a more negative attraction and the hydrogen portion has a more positive attraction. This results in water molecules aligning themselves; the oxygen of one bonds to the hydrogen of the next in a network of weak connections—so that instead of being a loose mixture of individual molecules, water is a latticework that behaves like a much bigger structure.

This little feature has tremendous significance. Without such hydrogen bonding, water would be like all the other molecules of its size and weight, a gas at earthly temperatures. Repercussions of hydrogen bonding (allowing water to be a liquid and act as if it were a bigger molecule) are why your veins are filled with fluid instead of gas.

Oddly enough, watery thoughts had also occupied John Dobson, who had flown in from the West Coast to join me for the "astronomy weekend" being held at the magnificent Victorian landmark of Mohonk Mountain House (in the Shawangunk Mountains of upstate New York), a mecca for rock climbers.

Dobson—whose "Dobsonian" telescope design has become the most prevalent in amateur circles in the past two decades—thinks we're even more intimately involved with water than most believe. A former monk of an Eastern order, now approaching age eighty, he lives in poverty deliberately rather than, like many of us, as a mere offhanded consequence of taxes or hedonism. A sprightly and

Two hydrogen atoms (lower corners) bind with one oxygen atom (black and white balls in lower center) at an angle of 105 degrees, giving water molecules a slight electromagnetic polarity. This is why water is not a gas at room temperature.

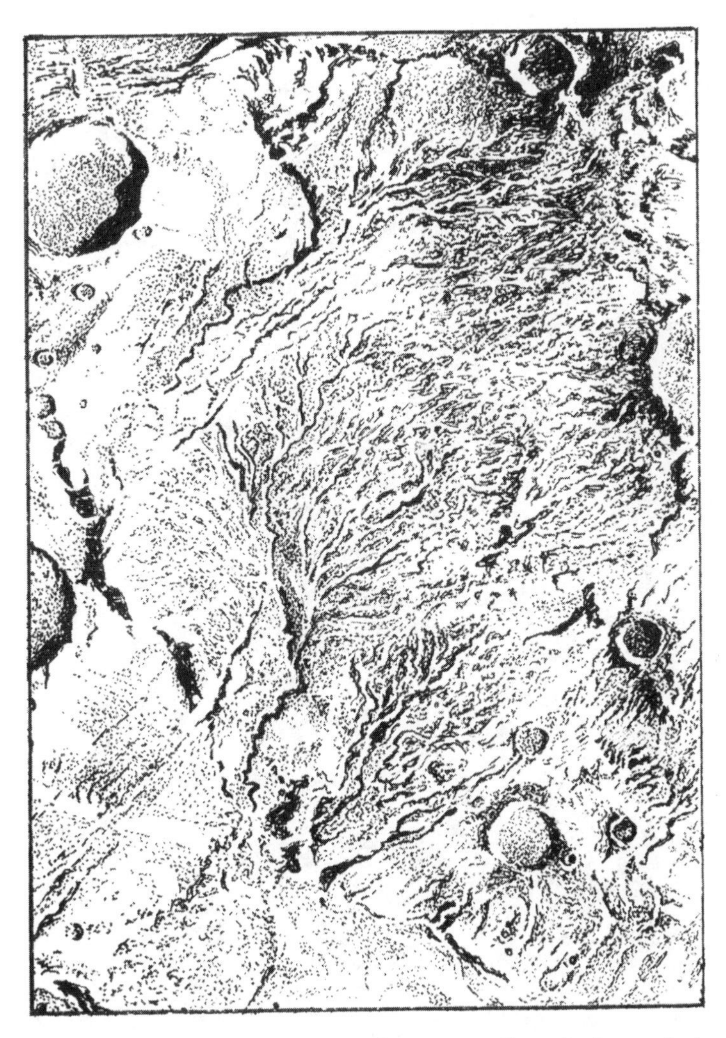

No water exists on Mars now, yet many Martian valleys display typical water drainage patterns.

original personality, Dobson sprinkles his lectures with digressing monologues about human origins.

"One night when I slept in my largest telescope," he recalls, "the peculiar acoustics and womblike darkness of the tube inspired a kind of clarity of thought about how our species originated." And now he is convinced that the genesis of the entire human race can be explained *aquatically*.

We didn't start out on the savannah as just a smarter variety of ape, goes this line of reasoning, an idea espoused in several of Elaine Morgan's books including *The Scars of Evolution*. Instead, our ancestors were stranded (probably in East Africa) during a period of rising sea levels. One large colony of apes became marooned on an island and had to learn to live on the beach. Before too long they made their peace with the ocean.

These human ancestors started using tools because they needed

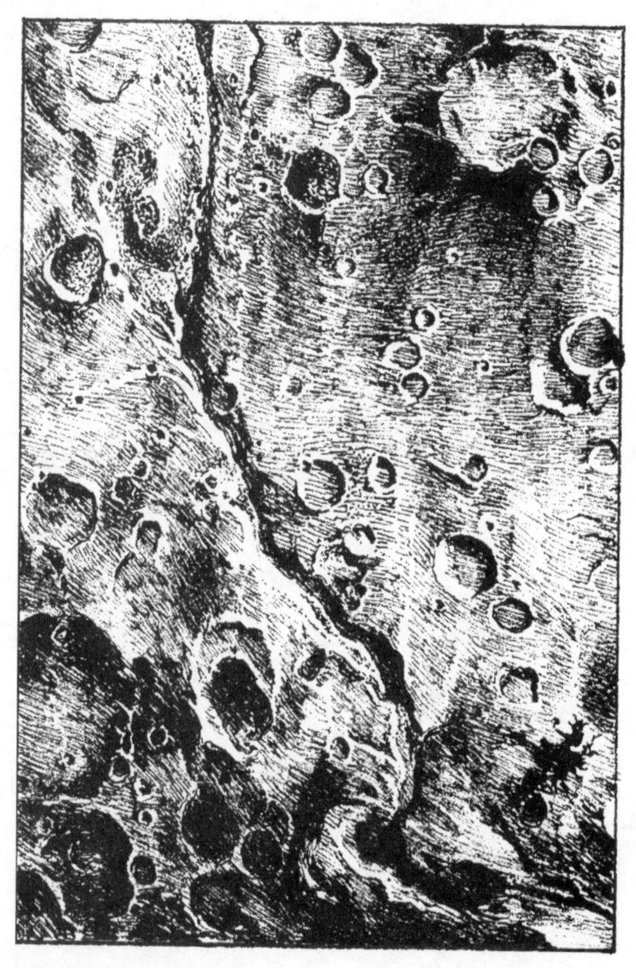

An ancient riverbed cuts through the Martian landscape. Something radically changed that planet's environment.

to pry open clams and such. They spent more time in the sea and soon lost their furry coats. Such an aquatic lineage explains our hairlessness; like dolphins, whales, and other aquatic mammals, we shed our fur because it is an unnecessary impediment in tropical brine.

Perhaps the hair on our heads remained so that our young could have something to hold on to when we swam. Our noses grew so we could breathe more easily when trying to keep our heads up. Our fat became attached below the skin, just as it is in dolphins and whales, rather than forming a separate layer, as it does in all the other apes and land mammals.

When surprised or terrified, we gasp. Why do we do this? Apes never gasp. Why do *we* gasp? It doesn't make sense—unless it's the vestigial legacy of a time when we took a quick, sudden breath in order to dive.

And we do indeed (except for pilots) love water. Other apes will cross water only when they have to, or if there's food on an opposite bank. But they don't love it. We do: We vacation on lakesides and by the sea, and they say a newborn baby will not drown if thrown into water. (At least that's what is claimed. I doubt anyone doing it would earn a "parent of the year" award. And onlookers would probably display that uniquely human gasp.)

While the aquatic-ape theory has yet to be widely accepted, I'd wager that school children a century hence will be taught precisely that explanation of our origins—an appropriate one, in a watery universe.

If our lineage, our lives, our brains, our food, our recreation, and nearly everything else we care about involve water, which in turn is just one of hydrogen's designs, it's clear that the universe did not have to consist of billions of disparate elements. There is creativity enough from just one or (including oxygen) two building blocks. Indeed, throw in yet another—carbon—and you've now got complexity beyond comprehension, including the ingredients for the amino acids that comprise all forms of life.

Since we ourselves are composed of these substances with their ever-changing forms, it is only logical that humans should manifest this quality of changeability. It's certainly not far-fetched to propose that anything so woven into the fabric of nature and inherent in our bodies will spill over into our personalities.

We don't ponder our own changeability: It's too close to see clearly. We do love changeability when we go from being grumpy to feeling playful. We dislike it when we go from health to sickness. When we're capricious and unpredictable, we don't ascribe changeability itself as the cause. And while each day carries a thousand myriad mutating moments, we focus on the things rather than the process, the flow. Ignoring our innate, watery, hydrogeny tendency to mutate, we vainly attempt to confine or control the process.

Can the rate of transformation vary? Can we change change? Not likely: Nothing inherent can be readily altered, and change is inherent. We try to preserve our youthful appearance and our possessions when a less frustrating viewpoint might let us enjoy (or at least accept) change in its own right. Then we might sit back as if at a theater and watch the show. This works, of course, only if we trust that the fact of changeability is more fascinating than any particular momentary manifestation, that the enchantment of the current moment lingers at full intensity only when we allow it to become something else. Attempting to preserve the ephemeral yields such mockeries as mounted deer heads and artificial flowers.

But what if the *rate* of change were itself variable? After all, red dwarfs go through their life cycles in slow motion and die at the Social Security–bankrupting age of 20 billion years. Blue stars speed through their stages a thousand times faster. Both are simply balls of hydrogen fusing into heavier elements; only the pace is different.

Imagine if the rate of change similarly varied here on Earth, if warm ice, say, suddenly refused to melt or heating a cup of coffee required months? According to Eastern thinking, the subjective perception of change does indeed alter with the individual's temperament.

(This is why I greatly admire Hindu and Buddhist thinking,

which has been obsessed with such intriguing philosophical issues for millennia and so has dealt with things like states of awareness and the paradoxes of free will—matters largely ignored in Western spiritual circles, which often limit themselves to the themes of faith and obedience.)

One's amount of *energy* seems to influence one's perception of time, but in a way opposite to what you might imagine. The more energy you have, the more exciting, varied, and rich life will appear to be and the more experiences and therefore hours there seem to be in a day. Children with their boundless energy feel as if it takes forever for them to pass through the childhood years of four to twelve. But a person in his or her sixties, now more sluggish, feels that the same eight years whiz by unforgivingly.

So perhaps changeability (or at least our perception of it) is changeable after all. In any event, our particular corner of the cosmos exemplifies and reflects nature's mutability to the hilt. As we watch our tragedies and comedies alternate—as our beautiful cities decay and then are razed and redeveloped, or, like Pope Gregory, we are sinners and become saintly; or we stroll through forests that were once farmland after standing as even earlier forests—our lives whirl through an almost disconcerting milieu of change. Perhaps that's why the stars, whose even greater fury is disguised by distance, seem so reassuring as they return each year—spurious symbols of stability.

We'd do well to follow hydrogen's simple creative examples of water and sunlight, ancient templates for adventures both here and probably in myriad other solar systems. Doing so, we submit to life's kaleidoscopic parade of new and wonderful designs—like cloud patterns that forever capture the attention of children and lure the astronomer and aviator to the sky.

Your Number's Up

Astronomy is the science most involved with enormous numerals, where numbers get tossed around as though one were haggling in a Turkish bazaar. Despite the fact that there are 30 trillion cells in the human body (far more than the number of stars and planets in our galaxy), human physiology is rarely associated with vastness. Yet expressing the distances between celestial points of interest is a whole 'nother story: Even using giant units like light-years brings us to bewildering numerals containing enough zeros to recall those I got in college German.

It wasn't always this way. The word *million* didn't come into general use until the thirteenth century. Before then, the largest number was a *myriad*, equal to ten thousand. The Greeks, who coined the term, would occasionally resort to myriads of myriads, and that was sufficient to express the most complex concepts.

A million seemed huge when we were kids. It became less intimidating only when we realized it was possible to count to a million in a few days. It's really not so big: A million steps takes you from San Francisco to Sacramento. A vacation lasting a million seconds gives you only an eleven-day reprieve from office misadventures.

In astronomy, we use *million* mainly in relation to the sun, which is nearly a million miles wide and sits 93 million miles away. *Million* also expresses the number of miles to the nearer planets. Venus is 26 such units, Mars 34. That's about it.

Even less useful, astronomically, is the billion, which is a thousand million (in the United States, that is; in England it's a million million). We might say that Saturn is nearly a billion miles away from us, and that Uranus, Neptune, and Pluto are a couple of billion. And the visible universe offers for our inspection about 50 billion galaxies. But that's where a billion's usefulness ends. No star has a radius as big as a billion miles, and none is as close to us as a billion miles. The unit just isn't very applicable much beyond Earth, although it's convenient for taking our planet's worldwide census of 6 billion people, some of whom may contemplate the accumulated wisdom of the 60 billion to 100 billion people who have ever walked the face of this forgiving planet.

So we jump to a trillion. This is a million millions, suddenly a most valuable unit for government economists, physicists, astronomers, and more than a few fanatics caught in the grip of computer nanospeak. There are almost a trillion stars in our galaxy, and about the same number of planets. The light-year is equal to 6 trillion miles. Grasping what a trillion represents is like having a floodlight illuminate the path to understanding the cosmos (and in comprehending our national debt—some $5 trillion).

One way to appreciate the enormity of a trillion is to count it out. Unfortunately, at the rate of five numbers a second, without stopping to eat or sleep, this exercise would still require 3,000 years. So a trillion seconds ago carries us back 31,000 years—to well before the dawn of recorded history. The problem with such numbers is that their zeros blur into incomprehensibility. For example, 1,000,000,000,000 looks not much different from 1,000,000,000,000,000. That is, a trillion resembles a quadrillion, more or less. Actually, they differ to the same degree that the weight of a tadpole differs from that of an elephant or that your body's mass compares with that of three cement trucks.

This is where we come to the realm more common to hyperbole than science. *I've told you a million times to clean up your room* leads children to think it's okay to exaggerate. Who, then, would object when textbooks continue to repeat the long-obsolete figure that our Milky Way contains "100 billion" stars—when the actual figure is four to ten times greater?

(It's easy to be imprecise with technological vocabulary. Case in point: People often use *silicon, silicone,* and *silica* interchangeably. But the meaning of *silicon* is elemental. And California's computer center is called Silicon Valley, not silicone valley—though some women who work there have probably utilized the latter for improvements cosmetic rather than cosmic.)

A few centuries ago, when it seemed as if the universe had somehow organized itself in sync with human conceptions, including our own arbitrary numbering systems, many scientists desperately tried to make planets' orbits and physical features match various numerological schemes. With enough searching, one that worked was sure to turn up—and it proved a real goody when it was uncovered by Daniel Titius in 1766 and popularized by Johann Bode a few years later. Eventually this layout became known as the Titius-Bode law.

Nowadays we don't call it a law because it has no physical basis for existing at all. But back then, because the coincidence seemed so incredibly compelling, it was assumed that a scientific justification for that arrangement would eventually be found. It never happened, and, okay, here's what we're talking about:

Start with the numbers 0, 3, 6, 12, 24 . . . where each number after zero doubles the preceding one. Then add 4 to each number and divide by 10 and you've got .4, .7, 1.0, 1.6, 2.8, 5.2, 10.0, and so on. What's so odd about this? Just compare it with the distances from the sun to the planets, expressed in Earth–sun spans (the unit astronomers use most often, called the Astronomical Unit, or A.U.).

PLANET	BODE-TITIUS	ACTUAL DISTANCE IN A.U.
Mercury	.4	.4
Venus	.7	.7
Earth	1.0	1.0
Mars	1.6	1.52
Asteroids	2.8	2.8
Jupiter	5.2	5.2
Saturn	10.0	9.54
Uranus	19.6	19.18
Pluto	38.8	39.44

Yes, this omits Neptune from the picture, which at 30 A.U. doesn't fit into this scheme at all, but Neptune wasn't discovered until much later. However, before that, when Uranus and the first asteroids were found, and they fit the pattern perfectly, this was interpreted as further proof that the heavens did indeed march in sync with the tidy numbering scheme. What the arrangement really proved to be was a giant coincidence, albeit a remarkable one. And instructive. For it teaches us to be very cautious when we attempt to link events with other events or with numbers schemes; it also helps us understand the difference between correspondence and coincidence.

Many of us have yet to learn that lesson. If old Aunt Lucy dies just as lightning strikes the church and stops its clock, we'd be sorely tempted to connect the two events and assign mystical significance to the whole thing. We forget that simultaneous events occur *all the time,* and just because the wind knocks down that big oak tree just as we turn on the cold-water faucet to brush our teeth, we mustn't assume that one event caused the other or that they're linked in any way. The distinction between correspondence and coincidence is one of the most critical not only in astronomy but in all the sciences, even if our minds have a much harder time making that vital separation.

When math is reduced to pure numbers independent of linkage with events or phenomena such as distances to planets, then coincidences are more easily seen for what they are. For example, the number 37 is prime, meaning it's an oddball, divisible only by itself and the number 1. Yet the numbers 111, 222, 333, 444, 555, 666, 777, 888, and 999 are *all* divisible by 37 and by nothing else.

Like Peter the Great, who had his wife's lover beheaded and kept that head in a bottle of alcohol in her bedroom for her to contemplate, nature can also be perverse, though usually with greater subtlety. There's no rhyme or reason for the numbers that biology and astronomy spring on us. Why does each cell in our body have 90 trillion atoms, roughly the same as the number of stars in our home cluster (Virgo Group) of galaxies? Why is there exactly the same number of Earth–sun distances in a light-year as there are inches to the mile? Why does the diameter of the sun (864,000 miles) have the same numerals as the number of seconds in a day (86,400)? We could go on and on, for these strange connections are always interesting. To the rationalist, coincidences are part of life; to the mystic, for whom everything is interconnected, coincidences simply do not exist. Both will extract immense but very different pleasure from matchups involving numbers large and tiny.

In any event, the trillion is the largest numeral we ever need to comprehend, for earthly as well as celestial use. A quadrillion—a thousand trillion—crosses the line so far beyond what the mind can handle, it becomes necessary to break it up into more digestible units. Like a giant chocolate bar, it's more easily absorbed in smaller pieces.

Astronomers oblige by slicing segments of space into light-years—units containing 6 trillion miles. The nearest star is 4.3 light-years, or 26 trillion miles, and so far the spin of our heads is still controllable. If we picture traveling from New York to California sixty times in one second—a conceivable exercise—then we have grasped the speed of light. If we imagine moving that fast for two months, then we have understood the distance of 1 trillion miles. Going that fast for a year means 1 light-year, which, you'll recall, is 6 trillion miles. A mental stretch, to be sure, but I think we can all do it and be left with some kind of a handle on the concepts of *trillion* and *light-year*.

But if we then insist on taking the trek farther, to, say, the nearest spiral galaxy—Andromeda—then we simply lose it all, especially if that gap is expressed as 12 million trillion miles, or 12 quintillion miles. That's meaningless. Much better to abandon the smaller unit of *mile* altogether (even if it is a familiar concept) and use the more esoteric *light-year* from here on out. Then we simply say Andromeda is 2 million light-years away, and whether we can honestly grasp that or not, we're at least playing with all our marbles.

The farthest distance we could possibly travel? It's a hitchhike from here to the edge of the observable universe. A mere 14 billion light-years. That may be conceptualized by some of us, but I'll bet that very few could make any sense or headway with that distance if it were expressed as 100 billion trillion miles, or 100 sextillion—a 1 followed by twenty-three zeros. Totally off the chart. We might as well use inches.

Still, those zero-rich numbers are tiny compared with the largest number of *things* in the universe—the sum total of all subatomic particles such as electrons. That figure is a 1 followed by eighty-one zeroes. Even if you counted electrons at the rate of a trillion a second starting when the present universe was born, you couldn't come close to tallying them. Yet, despite such vastness, any concept involving *things* has a limit, since the total mass of the universe is reasonably well known and is finite. If all the atomic nuclei of the universe, plus every electron, could be compressed so that no space remained between them, the entire universe would be easily contained within our solar system. Squeeze out the emptiness within and between each atom, and you're left with a ball weighing 10 followed by fifty zeros tons. Placed in the sun's present position, its surface wouldn't quite stretch to the planet Jupiter.

In other words, the actual material in the universe is infinites-

In the days before calculators, Sir Isaac Newton worked out many methods for dealing with large numbers. One recurring problem was multiplying binomials by themselves several times, for example, $(x + y)^4$. Newton reduced the problem to adding up a series of smaller multiplication products, such as $x^4 + x^3y + x^2y^2 + xy^3 + y^4$, but each of these terms needed to be multipled by a coefficient, and this is the interesting part. Newton determined that the coefficients would look like this:

$$1 + \frac{n}{1!} + \frac{n(n-1)}{2!} + \frac{n(n-1)(n-2)}{3!} + \frac{n(n-1)(n-2)(n-3)}{4!} \cdots \frac{n!}{n!}$$

The denominators are factorials ($3! = 1 \times 2 \times 3$) increasing step by step to the exponent factorial, while the numerators are factorials counting backward from the exponent. When both are factorials of the whole exponent, they cancel each other out. For the exponent 4, the coefficients are $1 + 4 + 6 + 4 + 1$. Hence $(x + y)^4 = x^4 + 4x^3y + 6x^2y^2 + 4xy^3 + y^4$.

Later, Blaise Pascal put these coefficients in a pyramid and found another relationship. His pyramid is called Pascal's triangle.

exponent	coefficients
0	1
1	1 1
2	1 2 1
3	1 3 3 1
4	1 4 6 4 1
5	1 5 10 10 5 1
6	1 6 15 20 15 6 1
7	1 7 21 35 35 21 7 1
etc.	

Each coefficient is the sum of the two numbers adjacent to it in the row above. If you have this diagram worked out to as high an exponent as you'll ever use, it can save the work of doing all those factorials.

Another interesting relationship arises if we add up the numbers that lie diagonally in the triangle.

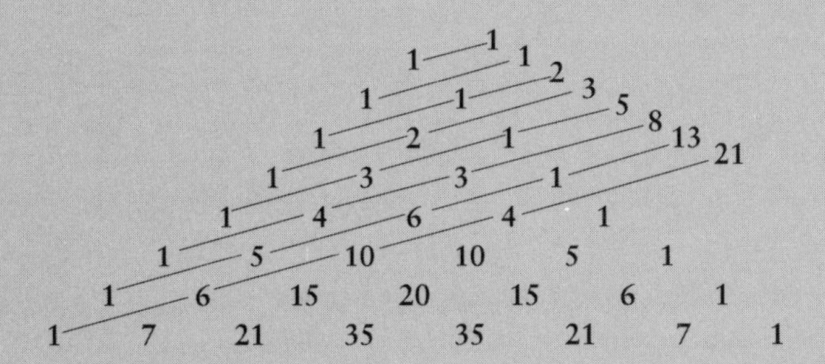

This new series of numbers is the Fibonacci series that turns up in golden ratios, spirals of seashells, the leafing of tree branches, pinecones, sunflower seed heads, and multiplying rabbits.

Since you've probably forgotten your high school algebra, we'll define our terms:

BINOMIAL: A number equal to the sum or difference of two other numbers (e.g., $x + y$).

COEFFICIENT: The number by which another number is multiplied (e.g., the 2 in the product 2y).

DENOMINATOR: The number you divide by (e.g., the 2 in $\frac{x}{2}$).

NUMERATOR: The number to be divided (e.g., the x in $\frac{x}{2}$).

EXPONENT: The power to which a number is raised (e.g., the 3 in y^3).

imally tiny when compared with the size or vastness of the cosmos. If the universe were a cube ten miles wide, ten miles long, and ten miles high, all the mass it contained, including even the mysterious dark matter, would be as a single grain of sand.

Here is how a googol looks in eight-point type:

10,000,000,000,000,000,000,000,000,000,000,000,000,000,000,000,000,000,-
000,000,000,000,000,000,000,000,000,000,000,000,000,000,000,000,000

On a typical computer you could type 4,536 zeros per page, so that it would take 221 pages of zeros to type out 10^{10^6} (1 followed by a million zeros). It would take 220,459 pages to type out a 1 followed by a trillion zeros. How many pages to type out the number represented by the word *googolplex*? Answer: 2.2 times 10^{96} pages, over a quadrillion times more pages than there are subatomic particles in the visible universe.

Yet even this, the number of subatomic particles in the entire universe, winds up being ten million trillion times less than a googol—a 1 followed by one hundred zeros—the largest existing number. (Unless you include the humorously contrived and almost infinitely larger googolplex, which is a 1 followed by a googol of zeros.) A googol can have meaning only in possibilities or permutations, since it far surpasses the number of possible *things* in the entire universe.

The reason possibilities will always involve far vaster numbers than things is this: To calculate permutations we have to use *factorials*. Let me explain. The number of ways four books can be arranged on a shelf is 4 factorial, meaning 4 times 3 times 2 times 1, or 24. (From this point on, let's ignore that "times 1" part of the formula.) Add just one more book, and the possibilities become $5 \times 4 \times 3 \times 2$, or 120. With a mere ten books, it's 10 factorial, or $10 \times 9 \times 8 \times 7 \times 6 \times 5 \times 4 \times 3 \times 2$, or 3,628,800! You see how the number of permutations gets out of hand very quickly. Now, what if the trillion cells of the human brain could each interact with any other? How many different connections would be possible? The answer is 1 trillion factorial, or 1,000,000,000,000 times 999,999,999,999 times 999,999,999,998 times

999,999,999,997 . . . and on and on. This figure is almost infinitely greater than the number of subatomic particles in the universe, and some 10 billion times greater than a googol.

Now, what if each of the electrons in the universe could "communicate" with any of the others? How many possibilities then? You'd have to perform 10^{81} factorial. The answer would be inconceivably vaster than a googol (but much, much less than a googolplex).

The purpose of all this is to make the lowly light-year, and the number of stars in our galaxy (a paltry trillion), and the distances from any point A to any point B seem very easy and workable when expressed in light-years.

Yes, astronomical sums are manageable. Our materialistic minds can come to grips with any manner of *things*. But when it comes to the myriad possible routes to discovering the universe, let us not count the *ways*.

View from a Window Seat

Can a long business trip be converted into a scientific adventure, an odyssey of colorful discoveries? In 1996 I found the answer: definitely. It involved little more than spending an extra half minute on the phone when making airplane reservations.

HarperCollins, the paperback publisher for my first book, sent me on a whirlwind seven-city publicity tour—endless television and radio appearances and day after day of jet travel between points A, B, C, and where-am-I-now. The low point came at a book signing. Having spoken with other authors, I was aware that book signing is pretty universally a dreaded, hated activity. Not because of facing the public; no, we all love meeting people and interacting with them for a few moments before signing the book we're happy they've bought.

The abominable part is sitting there like meat on a hook, looking dazedly off into the distance during those long minutes when nobody comes to present your book to be autographed. You feel awkward, a failure, and desperately hope more people will show up, and fast.

One day, after a lively, exceptionally successful TV appearance in a major Southern city, I was scheduled for the largest bookstore in that area. As I approached, I saw a double line stretching clear around a corner! Could these be crowds who had seen the TV show and now were excitedly awaiting my appearance?

Not quite. Turns out that at the last moment they'd gotten Greg Louganis, the Olympic gold-medalist diver, to sign his just-released autobiography. His struggle with AIDS had been on the national news all that week, and the line for his signature was ten deep for hours. Meanwhile, hiding behind neat piles of my suddenly ignored *Secrets of the Night Sky,* I sat at my own table, as miserable as a cat in a bathtub.

That night I peeked into the cockpit to chat with the crew of the jet bound for the next tour city, as I often do. Hearing that I'm a private pilot, they warmed to me and asked what I was in town for. I mumbled about being an author on a book tour. The captain asked the title of the book, then dropped his jaw.

"No! You're not Bob Berman!" he shouted; he had read and really enjoyed the book.* Suddenly the sting of the afternoon was gone. Until the next bookstore.

It was on that same flight that I realized how many truly amazing experiences are available to the jet passenger. As we enter the second millennium, the greater marvel might not be the *fact* of flight but its ready accessibility. That everyone can fly, anytime and anywhere, in the same lifetime as the airplane's invention, is a persuasive display of technology's democratic tentacles.

But the marvel seems to have lost its fascination. Restless, chronically bored airport crowds exude no sense of novelty or wonder, give no hint that they are about to do what eight thousand generations of their ancestors could not. Even passengers who love nature rarely choose their seat with deliberation, probably unaware

*In early 1998, *Secrets* led to my appearing on the *Today* and David Letterman shows. Every author's fantasy come true!

that they could, from 40,000 feet, gain glimpses of astounding phenomena that simply cannot be seen elsewhere.

So I'd urge everyone, after the standard expeditions to the washroom and magazine rack, or a glance at the second-run movie (which these days is shown in a special ultra-low-resolution video format), to consider a novel mode of entertainment. Open the window shade.

What an idea. Except that the blinding sun allows little to be seen other than a wing. So—let's start from scratch and *plan* the flight to harvest the sights uniquely available from seven miles up. It's worthwhile: Some of the finest gifts of the sky can only be viewed by those who have entered it.

A window seat, of course. You'll want it either in the very front or at the rear. In a 747, the wing can be avoided by choosing row 40 or beyond. In a 737, you'd want to stay away from 8 through 17. Choose the first ten rows or final five in a 727.

Next, you *don't* want the sun on your side. If the flight is west–east, choose the *left* side of the plane—seat A in most aircraft. Conversely, of course, an east–west route dictates the right side. In a north–south flight, time of day becomes important, since the sun will be on the left only in the morning.

Use the time while the plane is taxiing to assess the sky. If it's partly or mostly cloudy, are they low, puffy (cumulus) clouds with patches of blue in between? Or a flat layer (stratiform) with breaks? Either way, as long as there isn't a solid layer of clouds above them, the stage has been set for an intriguing and colorful apparition.

As soon as the jet completes its climb through the clouds, look downward—for the *glory*! Seen only on cloud tops from above, the glory is a colorful series of concentric rings that surrounds the spot exactly opposite the sun. That antisolar point requires no great search: It's marked by the plane's shadow.

Moreover, the vivid rings of color are explicitly centered on the part of the shadow that corresponds to your seat! If you're in the tail, then the shadow of the plane's tail is the center of the glory.

Caused by refractions from tiny droplets, the glory's size can

Just after ascending above the clouds, a plane casts a large shadow surrounded by an intense glory. *As the jet climbs higher, the shadow becomes smaller; the glory remains the same size but grows dimmer.*

vary enormously as it floats like a phantom on the upper side of the cloud layer. This itself is curious, since other refraction phenomena such as rainbows and halos come in only one size (with just one additional, dimmer secondary arc sometimes seen). The glory's dimensions reveal the nature of the droplets in the clouds below. The larger the glory, the smaller the raindrops.

Do not give in to the impulse to tell the person sitting next to you, "I see the glory!"

But do contemplate what exactly is going on, and why you've never seen this phenomenon before. Remember that the halos are centered on a cloudy spot opposite the sun from you: When, in everyday life, does such a thing happen? On Earth, the spot opposite the sun from you will never be a cloud, it'll be the ground just below you; specifically, the antisolar point is always marked by the shadow of your head.

Since sunlight is always bounced straight back from that spot in the same way the light from a projector is reflected by a movie screen, you can often see a bright glow, like a halo, surrounding your head when you look at a dewy lawn, as if you've suddenly and undeservedly attained sainthood. But you really need a wall of fog or cloud in that position to get the multiple reflections and refractions of a glory.

Its nearest relative on Earth is the *brocken bow,* a phenomenon long seen in the Alps but available from any mountaintop under the unlikely circumstances of your looking downward onto a nearby cloud that's also opposite the sun.

A jet, during the usually brief period when it is in sunlight just above the cloud tops, provides the right ingredients routinely. The glory's beauty and effortless availability make it a rewarding quarry: How sad that so few passengers know to seek it.

As the aircraft climbs and its shadow on the clouds below shrinks, the glory fades, the sky darkens, and the horizon drops. In effect, you are increasingly peeking over the curve of Earth to ever-greater distances. The moon or sun will appear even if it is beneath the horizon for people on the ground below, and a low sun or moon will clearly float *below* you, more than 90 degrees from the zenith.

At full cruising height, one is treated to a miniature version of an astronaut's perspective: more than 180 degrees of sky. Actually, astronauts have a more limited panorama than is commonly realized. They orbit no higher than 350 miles, or one twenty-fifth of our planet's diameter above the surface. Relatively speaking, they are barely above the Earth. The strongly curved horizon depicted in live TV shots from space is an artifact of the extreme wide-angle lenses of their cameras. The dramatic curve of Earth is a fraud.

As you gaze out the window, a most basic question arises: Just how far can you see? The formula for figuring the distance to the horizon is so uncomplicated, it can be instantly memorized. It's as easy as one-two-three. Namely: 1.23 times the square root of your altitude gives you the horizon's distance in miles.

It's not as complicated as it sounds. For example, say you're sitting on a beach chair at the water's edge, your eyes some four feet above sea level. Square root of four is two, times 1.23 yields 2.46, and there you have it: The horizon is about 2 1/2 miles away. No wonder it's such an obvious sharp line out there at sea: It's surprisingly nearby! Any small object farther than 2.46 miles away is out of sight over the Earth's curve.

At 40,000 feet, the arithmetic is almost as simple. That number's square root is a gratifyingly even 200, times 1.23 becomes 246. So you're 246 miles from your horizon.

This means that at cruising height, you're centered on a visual circle that takes in a vast 500-mile stretch, embracing some 190,000 square miles. That's about 7 percent of the entire contiguous United States: Always within view is the equivalent of Austria and Switzerland combined!

Come nightfall, you'll see luminous dots outside your window. Above 75 percent of the Earth's atmosphere, the sky is noticeably darker than on the ground, and there would be 30 percent more stars visible were it not for the layers of glass and Plexiglas standing between you and breathing difficulties. If the moon is visible, it will appear larger, but not because you're closer to it. Actually, unless it's more than halfway up the sky (in which case it becomes increasingly difficult to see because your face must be pressed against the glass), it's no nearer than when you were on the ground.

The moon's larger size is an illusion caused by its being framed by a small window. Even if the moon were overhead, you'd be closer to it only by one part in thirty thousand, producing a size increase that even the world's largest telescope could not measure.

Gazing downward, you see the effects of a century's electrification. The bluish dots are the mercury-vapor streetlamps, routinely used in the Third World and still common in the United States despite their energy inefficiency. The pinkish-yellow dots are the newer sodium-vapor lights, the current prevalence of which has transformed the aerial scene since the 1960's and is responsible for the amber glow above cities.

The jet engines' hum, generated by burning kerosene, can lull you into forgetting the passage of time. But contemplating time is a good way to spend it, observing how the interaction between moving jet and rotating planet alters the rhythm of day and night.

On the ground, the spinning Earth carries us at a rotational speed that varies with latitude. Got a calculator or love math? Then you can quickly calculate your cruising speed on our whirling planet, for it's simply the cosine of your latitude multiplied by

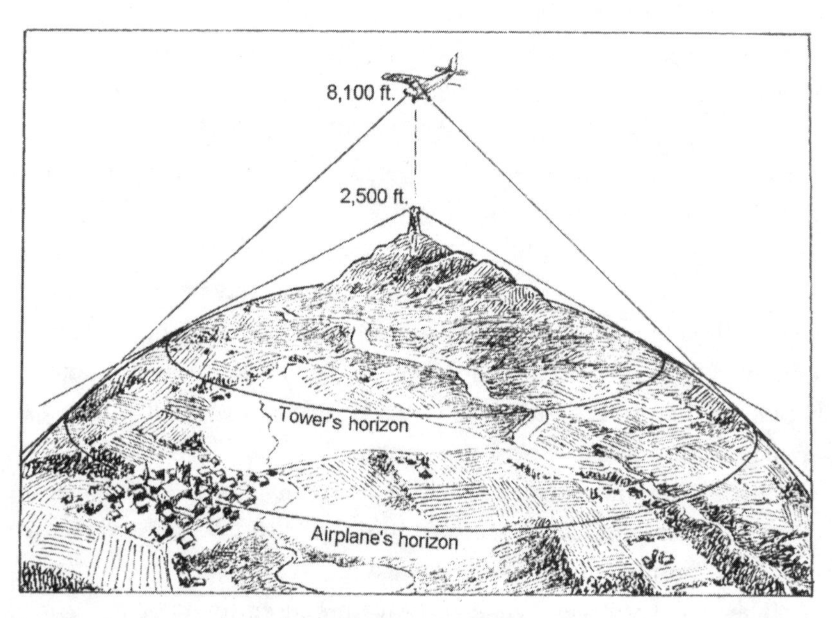

Can you see forever on a clear day? Flying at 8,100 feet, you'd see a town 110 miles away (lower left), *but the town is far beyond the horizon for the fire-watcher in the tower directly below the plane. From her 2,500-foot elevation, the horizon is located 61 miles away.*

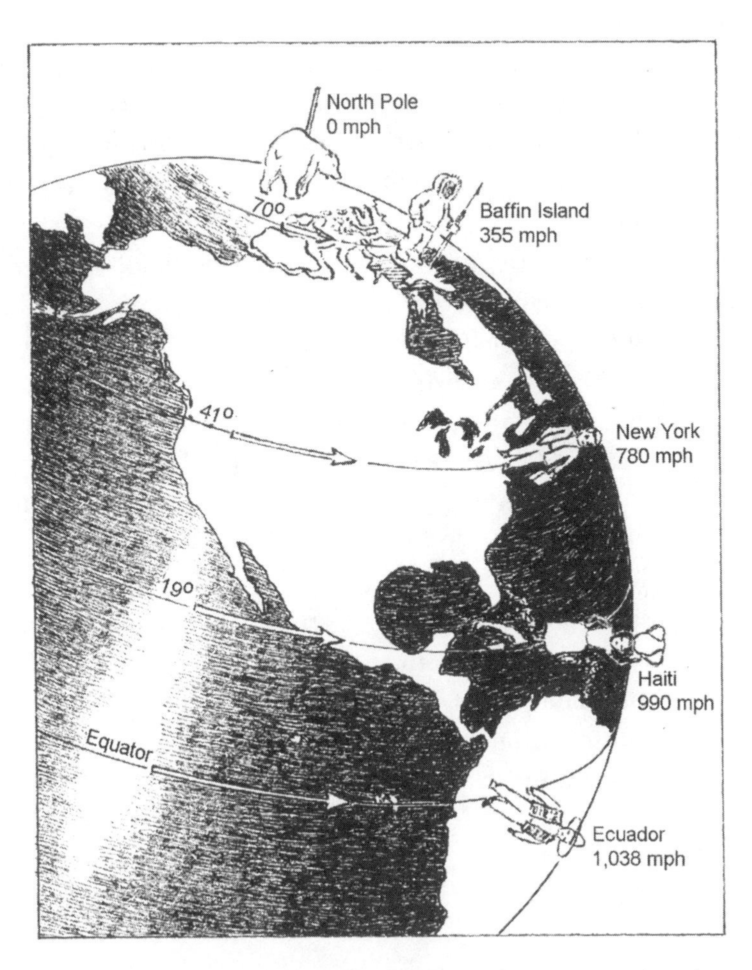

Location is everything. Your speed on our spinning planet is equal to the cosine of your latitude multiplied by 1,038 miles per hour—the equator's velocity.

1,038. But if words like *cosine* turn you off, just forget the whole thing; I've done the math for you:

People at the equator speed by at 1,038 miles per hour, while New Yorkers, for a change, shuffle along a bit more slowly, at about 780 miles per hour. Penguins at the pole are essentially stationary, performing a single leisurely pivot every twenty-three hours, fifty-six minutes.

It's come to pirouetting penguins to make this point: Your jet, tooling along at some 500 miles per hour, can't quite nullify the

Earth's rotation. That is, unless you're traveling at a greater latitude than 65 degrees or so, for example, on a polar route or passing over Alaska. In that case, the sun can indeed travel backward in the sky, or be seen to set and then rise again in the west like a funhouse mirror!

When you land in Australia or another destination solidly in the Southern Hemisphere, the moon and sun *always* move backward. The sun still rises in the east and sets in the west, but instead of getting there by moving to the right, the way Americans, Russians, Europeans, Japanese, and most of the world have seen it all their lives, it *travels leftward* throughout the day. Tourists there often *sense* that something is deeply strange but cannot immediately say why.

Flying west to east at high latitudes, the days and nights flash by in almost time-machine fashion. A full cycle of sun and darkness can be compressed into less than twelve hours. It's incredible, it's real-life science fiction, but it's got tough competition. Which wins out, nature observations and thought experiments, or the in-flight movie, *Nightmare on Main Street* . . . ?

Certainty Principle

To say someone is insecure is insulting. We all want to feel confident. From a child cuddling in a parent's lap to the retiree looking over a financial portfolio, certainty sits atop our list of desirables.

That all orthodox definitions of *security* correspond to ephemeral things and are thus ultimately insecure is a contradiction left to philosophical debates. When we examine nature's kingdom to see how security fares, we do so academically, with no intention to rethink our own stance. The canary may indeed chirp gaily despite not knowing if his seed cup will be refilled, just as Earth's biosphere continues its complex interrelated pageant untroubled that an incoming comet might even now be en route, destined to break its intricate designs into a thousand unsalvageable fragments.

Such carefree lifestyles may be fine for birds and galaxies, but not for you and me. (True, occasionally there comes a heretic like the late philosopher Alan Watts, who preaches *The Wisdom of Insecurity,* arguing that greater wisdom and joy come to those who live fully in the unknown and unknowable. But level heads must prevail: We're certainly not going to follow such dangerous, abstract counsel, at least not until we first see how the rest of nature fares.) There may or may not be balm in Gilead, but is there at least *something* reliable enough in the world of physics to serve as a template for our own need for security? Or instead—as in Chaos theory—is there little that can ultimately be predicted with certainty?

At first glance, the life cycles of stars and planets seem tediously and methodically dictated by their constituent atoms, which in turn appear born to march along the boulevards of immutable and trustworthy laws. Prior to this century, you'd have been hard-pressed to make a case that physical reality is in any way undependable. The successes of Newtonian mechanics, which let us predict even the existence of new worlds (Uranus in 1781 and then Neptune in 1846), showed that nature functions with Teutonic rigidity and unforgiving precision. No wiggle room, no doubt, no insecurity at all. And best of all, our minds—*our minds!*—could grasp these processes to perfection. The universe moved in accordance with canons that were precisely understood by mathematics and science even when we employed arbitrary numbering systems we ourselves invented.

It was so beautiful, so perfect, how could we not snuggle in an unshakable faith in our own intellect? Collectively, we all came to have a mind-set shared by nearly every educated person in Western societies. The communal sense of reality went something like this:

We already have much if not most of the universe within view. We certainly do not know everything. But everything is *potentially* knowable, and much of it *will* be known, in time.

This outlook seemed to embrace a reasonable mix of confidence and humility, and it offended no one. Even the Church found no fault with it.

Then came quantum theory.

Waging nothing less than total nuclear war on our ability to grasp the universe intellectually, it demonstrated that phenomenon after phenomenon could be explained only if things at the most basic levels of existence were utterly unknowable except in probabilistic terms.

Einstein hated it. He gave himself a migraine trying to belittle this new outlook, and he had lots of company. Far more than just an original way of explaining how nature works on the smallest scales, this theory paradoxically undermined our fundamental confidence in being able to grasp reality. While it allowed us to predict the behavior of groups of subatomic particles, it *altered* our view of the very nature of reality.

Quantum theory is explored later, but first let's pose a basic, underlying question: What, exactly, *is* security?

Since our homes represent in principle the epitome of security, we probably should leave them behind to consider the issue properly. That credit card commercial warning you, "Don't leave home without it" is right on the mark: It strikes a nerve because leaving home means stepping into the unknown. Do we ever know for sure what awaits beyond the front door?

(Over the past two decades I have nearly put my foot on a rattlesnake at the bottom step to my deck, glimpsed a bobcat crossing my driveway, and dealt with a flood that reduced my entire 500-foot dirt roadway to prehistoric muck—and those were merely events just in front of my front door.)

Traveling beyond familiar boundaries is security's antithesis: To mitigate the experience, many people arrange hotel reservations in a new city far in advance. I do it on a lecture tour or book promotion. But not on a vacation even to a totally unknown place. The prospect of finding myself condemned to a boring, boxy, chain-style motel suite, or else an independent hotel's dark room overlooking a brick wall or situated over an all-night disco, is not worth the security of a prearranged bed. Instead, four years of living overseas and many more years of finding my way in more than thirty-five countries have brought me to feel good about traveling "blind." If necessary, I ask a cab driver for recommendations: It rarely fails.

Going someplace without knowing where you'll wind up may

seem a trifling example of living insecurely, but many people feel too vulnerable to do even that. Now, this may seem like smug self-aggrandizement, holding myself up as the paradigm of fearless travel. But that's not the case: Since I derive security from first seeing my accommodations before agreeing to the room, I'm actually living less cavalierly than friends who will book sight unseen. In a sense, it is *they* who are taking risks, trusting to the unknown. Weighing both, I suppose that neither is a valid showcase of fearless living.

Does anyone in this comfort-loving, risk-avoiding age *ever* act with the here-and-now freedom of the animal kingdom? The venerable conflict between independence and security keeps arising, especially in relationships, but usually little real courage is demanded. If Moses stepped into the Red Sea knowing that God was on his side, that hardly took much nerve: You're not really sticking your neck out when the Ruler of Everything has lent you his credit card. But if Jehovah laid low for a while, so that Moses wasn't quite sure how things stood but then walked into the sea with hordes of irate Egyptians at his heels anyway, well, that's another story; such tests of faith are the traditional religious explanation (or rationalization, if you're a skeptic) for God's seeming absences.

As a motorcycle enthusiast, I biked for years in Asia without a helmet. Driving along the dangerous curvy roads of the Himalayan foothills near the Chinese border, I preferred my head to be pummeled by the swirling wind of a hundred smells rather than have the dark, confining safety of a brain-preserving helmet. Now, a quarter-century later and still a cyclist in the warm months, I wear the headgear, but in many states bikers are demanding the repeal of helmet laws so they can choose for themselves between hedonism and common sense.

If stars had free will, how many would go for the safe 30-billion-year life span of the feeble red dwarfs over the lightning intensity of the blue supergiants, doomed to death in a matter of a few million years?

"The star that burns twice as bright lives only half as long!"

explains the inventor Tyrell in the cult movie *Blade Runner,* when one of his powerful "replicants" demands to know why he had not been given a longer lifespan. (Apparently unsatisfied with the explanation, the cyborg then kills his creator.) But we face the same choices routinely and voluntarily. Live an exciting but dangerous life as a test pilot, for example, or vote for the extended actuarial life span accorded to librarians?

Since stars presumably have no say in the matter, what process makes such choices for them? Whatever, it obviously favors safety: Faint, long-lived red ones outnumber the others by twenty to one.

I knew an anthropology professor, born into an extremely wealthy family, who became so dedicated to a spiritual pursuit that she went to India, threw away her passport, and proceeded to live as a mendicant. Determined to accept whatever was to come her way and let her meals be provided by God alone (as she saw it), she gave away all her money and lived solely on alms. She's still there now, thirty years later. In the "living insecurely" department, she's the champion.

The strangest example I have encountered of a person exhibiting this characteristic came along in 1968, when I was twenty-three years old and met a beautiful redheaded Danish woman named Rekke. She very clearly was . . . well, different. For one thing, she smiled incessantly, a peaceful smile as if experiencing her surroundings with unrelenting innocence and joy. I was intrigued at how directly and clearly she'd look into one's eyes, at the amusement on her face, and also at the fact that she never spoke. Here was the Mona Lisa squared—mysterious and attractive beyond words.

We both lived in a small village overlooking the Nepalese countryside, and I pursued her avidly for a few weeks. Courting someone who didn't speak necessitated creative addenda to the dating rule book, and there was also this little complication: She appeared to be insane.

One example: She'd smile her way along the crowded temple-strewn boulevards . . . with her blouse flying open. She didn't expose her breasts deliberately; it was obvious that she simply didn't care enough about what anybody thought to bother buttoning properly. Naturally this would draw an accompanying crowd of appreciative Nepalese men awaiting the next breeze, while I'd entreat, "C'mon, Rekke. This isn't cool. Just button up. This isn't a European beach. You're in their ballpark; play by their customs." She'd ignore it all and keep strolling as if on the moon. Was she enlightened or just plain loony?

A few weeks into our relationship, I finally got her to speak. We'd hiked into the foothills and were sitting on the edge of a cliff. She was bright-eyed and radiant as always, and the valleys spread before us in the orange late-afternoon sun, thousands of feet below. The cliff face was not quite vertical but angled downward at perhaps a 70-degree drop.

We smiled brightly at each other. And then I jumped off the cliff.

This was a close call. I wasn't sure whether the cliff face departed sufficiently from vertical so that I'd simply roll and slide a bit before stopping, or whether it would be just too steep and I'd keep picking up speed until I hit the rocks at the bottom and kill myself. Close call, but I just did it.

When I scrambled back up, she rewarded me with the first words I ever heard her say. In a nutshell, she said (predictably) that she loved my fearlessness, my Zen-like leap, even though you and I know that it was moronic in the extreme—a new high or low in the annals of how dumb people can be when attempting to impress someone. Coming within a whisker of killing myself, I had reverted all the way back through evolution to something akin to the praying mantis.

All this is merely offered as background, a preamble to the rashest thing I ever saw her or anybody else do. We'd been walking through a village when Rekke silently handed a vegetable vendor a rupee, took some carrots, and started munching on them. "No," I advised, "you can't eat raw fruit in this country. You'll get dysentery."

Whereupon, without a moment's hesitation, she stopped and bent over a pile of dog feces we happened to be passing, scooped up a wad with her finger, and put it in her mouth.

She was eating dog excrement! What could I say? In a flash, she'd graphically demonstrated her faith and fearlessness. The immediacy of her action was startling. On the other hand, being a slow learner, I was no closer to determining whether or not she had all her marbles. And anyway, what difference? Was there a clear line separating sane people from the rest? So far as I knew, schizophrenics were troubled individuals, haunted by delusions. Rekke exuded serenity and joy. Could her Zen-like lifestyle somehow place her beyond the biological cause and effect of bacteria and disease pathology? (Apparently not. She came down with hepatitis when she got to Delhi. Which was about when we lost track of each other forever . . .)

But, to return to our theme, we lose track of people, but neither I nor any serious astronomer ever loses track of the planets. Any knowledgeable amateur can point toward each planet and do so even in a closed room. Some researchers have gone farther and wondered whether the planets will lose track of *themselves*. The question "Are the planets secure in their orbits?" has been the subject of much computer analysis. In the short run (the next ten million years), the solar system's integrity seems safe. After that, however, orbital oscillations caused by the incredibly complex gravitational interplay may well throw Mercury or other planets out of the solar system.

All right, and so what? We never really needed Mercury. We suffered no diminished quality of life as children before we first knew it existed. But in even longer time spans, other planets may prove to have ultimately unstable orbits as well. How many are expendable?

Earth and the larger planets seem particularly secure, but what if we weren't? What if we knew that, say, a mere million years from now we'd fly into the sun? Would it affect our attitude?

You bet. There's something reassuring about our present exaggerated and unjustified sense of safety: You hear it all the time from students. There's truth behind the old joke about the student who worriedly asks the teacher, "Did you just say that the sun will blow up in 5 million years?" And when told that it's 5 *billion*, he visibly relaxes: "Whew. For a minute I thought you said 5 *million*!"

But if it were just one million, or better still one thousand, we might live with the sense of immediacy experienced by the condemned. We might just share with our celestial boatload of fellow wanderers the cherishing of life known to those who feel in their bones that everything is, after all, wonderfully impermanent.

I wouldn't, as an astronomer, have to look far if I were the worrying type. The sun has a famous eleven-year cycle in which sunspot storms wax and wane. In recent decades it's become abundantly clear that the sun emits less energy (and Earth gets significantly colder) during the times of sunspot minimum. The return of the spots, occurring now near the turn of the century, is reassuring; our own healthful terrestrial biological patterns depend on it.

But something happened in the mid-seventeenth century. For a seventy-five year period from 1640 to 1715, the sunspot cycle simply stopped dead in its tracks. Simultaneously, bizarre global cold endured for that full human life span. Called the Maunder Minimum, the nearly complete loss of the sunspot cycle produced an icy period of extraordinary hardship and widespread suffering. The English Channel froze solid, and glaciers started heading southward. We had at least one "year without a summer." The big question is, what happened and why?

The answer from astrophysicists is, at least, refreshingly unambiguous. They say, "We have no clue." That miserable event ended less than three centuries ago, an eyeblink in the sun's life span. What security do we have that the present solar cycle won't be the last for another lifetime? And with it, any lakeside summer vacations?

No, there is no security in space, nor in the space between our ears. For even if we do discover why the sun is fickle, we're still utterly impotent to influence its behavior. All the technology at our command could not influence the sun in the slightest. We could not measurably alter a single solar characteristic even if we hurled our entire planet into its core.

Astronomy, unlike any other science, gets to examine only a frozen snapshot of what is really an almost infinitely complex, slow-motion ballet. We still do not know whether galaxies or stars came first, or how the immense superclusters of galaxies formed. Indeed, the existence of dark matter is inferred primarily because such conglomerations should be flying apart but aren't. Their security from dissolution causes our intellectual *insecurity,* since we have no idea what's holding them together. We call the presumed material dark because it's invisible, but it's really dark in the same sense as "darkest Africa": The light of our intellect can't reach it.

Sunspots appear dark because they are cooler than the surrounding gas, but they're still 5,400 degrees Fahrenheit and would be dazzling white if viewed amid different surroundings.

*The five bright galaxies in this image have similar red shifts and are assumed to be orbiting a common center of gravity, yet their speeds are too high to be controlled by the gravity of their visible mass: The cluster should have flown apart. Its continued existence conjures speculation about dark matter—unseen material whose gravity supplies the missing glue. (*Left to right, *galaxies IC 5359, IC 5356, IC 5357, PGC 72405, and IC 5351.)*

In the final analysis it makes no difference. Security, unlike love, is independent of nothing. Mad, beautiful Rekke once said, while gazing wide-eyed at the heavens, "Every day a *ball of fire* crosses the sky. Isn't anybody picking up on that?"

I followed her eyes from the sky down to the villagers. But even in rural Nepal, they were much too busy to look up, too occupied weaving individual parcels of security. What a mundane miracle it had all become.

Night Sweats

THE NIGHT SHOWS STARS
AND WOMEN IN A BETTER LIGHT.
—*Henry Wadsworth Longfellow*

You wake up in the middle of the night; there's a creaking noise from the next room. Your eyes strain at shadowy forms. A flitter of primal dread competes with a rational attempt to dismiss it as the wind, the cat, anything benign. You have experienced the most common and ancient fear—fear of the dark.

I see vivid demonstrations of this whenever a city friend comes up to my forest-encircled home for a visit. People accustomed to urban settings display an amusing paranoia about rural life, much as they would have centuries ago. They'll be apprehensive of the forest, wary of insects, nervous on unlit country roads. Taken for a walk on a moonless night, they're genuinely spooked.

Country residents often find the same enveloping darkness, with its myriad of stars, a source of peace and infinitude. *Their* foreboding comes when visiting a noisy metropolis with its ominous alleyways and abundance of weird, combative, or suspiciously friendly characters—"Lady, let me carry your suitcase for you." Given the choice of dangers, most rural residents would happily take their chances on a dark country road. Like a frantic, miserable cat transported in the family car, we're wary of any alien environment, natural or synthetic.

But haven't city dwellers always been correctly in touch with a fundamental instinct? Aren't humans genetically programmed to be leery of the night? Isn't the city a bastion against the dreaded darkness, that vestige of the time when menacing nocturnal creatures routinely produced the kind of anxiety we now seldom experience outside of parking space confrontations?

My own child is typical of juveniles everywhere, who find darkness particularly worrisome. Ghosts don't emerge until nightfall, and that shirt slung over a chair surrenders its monster shape only with the dawn. But even many adults retain a nocturnal discomfort. Is there anyone who hasn't heard or said, "Things will seem better in the morning?"

There's no shortage of ancient cultures that feared darkness, not only because of tigers and such, but because of their dread of ill-defined demons and unspecified threats from the night sky itself. And the strange netherworld of dreams does nothing to make night seem less supernatural. Despite modern "dream labs" and sleep experiments, the little we know is more mystifying than clarifying. No one can explain why, as night passes, dreams get longer. Or why early dreams relate to recent events while later ones tend to be more surreal and mysterious in content.

So where do these night sweats come from? Is the starry canopy inherently dangerous?

Some behavioral scientists speculate that it's not merely a vestigial fear of predators that programs us to be wary of darkness; they suggest that cataclysms such as widespread destruction caused by meteors, or the blackening of the sky by volcanic dust, have left their genetic fear-imprint. Rationally, such ideas seem far-fetched, since the sky is, and has always been, a rare and improbable source of human death and injury. Only one person in recorded history has ever been struck by a "falling star"—and that was by day. Mrs. E. Hullett Hodges, in Sylacauga, Alabama, suffered a bruised leg in 1954 after a meteor came through her roof and ricocheted off a radio. Many others (in 1992 and in 1996 in the United States, for example) have had close calls when meteors plowed into their lawns just as they walked by. Again, all by day.

But I'll confess: As it has for many, the most frightening moment of my life occurred in utter darkness. It was during a hurricane's torrential rain, when I was thirteen and my family lived in New York City. I visited a friend and was immediately warned by people in the lobby of his apartment house that the elevator was unsafe because the basement was seriously flooded. Naturally, I had to check out the situation for myself. With typical teenage bravado I stepped in and pushed the button marked "B." The elevator started downward easily enough but stopped abruptly with an ominous deep thud and splash. Then it happened: Inky water streamed in from the space under the door. As I watched in horror, black water rose first to my ankles, then up my legs, while I frantically pressed the button for the lobby. Then the lights went out.

The end to my terror, due to undeserved luck (or perhaps the fevered promises God knew I could not possibly keep), came when the car mercifully started upward, discharging its heavy load of water as the door opened in the lobby while startled adults leaped away from the sudden flood.

I've felt nighttime apprehension other times, as when flying my small plane on a rainy night. But generally, the nocturnal realm feels friendly. And why shouldn't it? Now that our forests are filled with nothing more bothersome than garden-destroying, tick-ridden deer, we can fearlessly take our pre-bedtime walk while the stars wheel benignly around the heavens.

Yes, we're usually safe within Earth's sheltering cocoon. But, as astronauts have repeatedly experienced, beyond our atmosphere lies a dark environment fiercely hostile to life: The most unforgiving Himalayan peaks are nursery-school friendly by comparison. Every space-walking astronaut knows that invisible apple seed–sized meteoroids constantly streak through the void. Their minimum speed, twenty times faster than high-velocity bullets, would puncture space suits like knives; bodies offer them barely more resistance than fog.

Astronauts don't need cyanide pills to provide a euthanasic exit in the event of an irreparable mishap. Every manned spacecraft is already immersed in a matrix rife with mechanisms for speedy death.

In space, the sun's searing ultraviolet rays can deliver a painful burn in ten seconds. A three-minute exposure would cook skin to carbon. The vacuum of outer space is worse than merely having no air; the pressure differential forces body gases outward: Eardrums and arteries would pop.

The sunward-facing side of the unprotected astronaut's body would rapidly heat like a microwave to 250 degrees Fahrenheit, above water's boiling point. Simultaneously, the other side would quickly freeze to 240 below zero, cold enough for skin to "crack" into solid ice.

All this would be the minimum effect, the kind encountered in the relatively friendly environs just above Earth. If the experience were to occur near Jupiter, then awesome, cyclotron-level radiation would sterilize all living tissue. Nearer to the sun or stars, bodies would immediately blister, char, and then vaporize. In the vicinity of white dwarfs or neutron stars, tidal effects would rip one's skeleton to pieces.

In most parts of the universe, far from anywhere and anything, the mechanism of death would simply be the cold, hard vacuum. A person, composed mostly of water, would revert to the commonest form of H_2O—a block of ice at a stable temperature of 454 degrees below zero. If subsequently struck by a fellow piece of celestial flotsam such as a meteoroid, the body would simply shatter into pieces, perhaps dividing itself along internal "fault structures" defined by organs and tissues.

The *Apollo 13* drama has been well chronicled, thanks to Jeffrey Kluger's superbly written account. Still, that four-day ordeal, in which three people shivered pathetically in a space as small as a couple of phone booths, is hard to imagine. (And the movie merely hinted at the harrowing mental and physical suffering involved.) Electricity shut off. Near-freezing blackness. Body moisture condensing onto wet metal walls. Human excrement floating surrealistically as drifting globs. With no warm clothing, and shivering uncontrollably, these three knew that their coming plunge through the atmosphere might well culminate in incineration—if their odd little craft didn't skip off into space never to return. Death by fire, ice, or suffocation. No middle ground. No temperance beyond Earth, not even in death.

By the time of American moon landings, fourteen Soviet and American astronauts had perished in spacecraft mishaps. Many others had had extremely close calls. Even Neil Armstrong, three years before his historic moon walk, nearly died when flying a Gemini capsule to the first-ever docking with another spacecraft.

Half an hour after the two had joined, the linked duo starting spinning uncontrollably. Both crew members assumed the problem had originated in the unmanned craft and fired a thruster to separate the pair—the worst thing they could have done, as it turned out. *Gemini 8* started tumbling ever faster. Armstrong urgently radioed Earth: "We're toppling end over end. We can't turn anything off!"

But the gyrations grew more violent. Soon ground controllers could receive neither telemetry nor intelligible voice contact. Both astronauts became incapacitated as the craft performed a complete spin each second. Their vision blurred; they could no longer read the dials and gauges and started feeling the first signs of blacking out. Amstrong played a final, desperate card. Breaking a mission command, he shut down all systems and used the controls reserved for reentry. Slowly the craft responded. A stuck maneuvering thruster had failed in the on position.

Two years later, Armstrong again cheated death when he bailed out of a training vehicle barely 200 feet above the ground, just before it crashed in flames.

But with all this, nobody has ever actually died above Earth's atmosphere, let alone had his body become irretrievable. Such terrors surely await future manned spaceflight.

The dark horrors can run deep. If we ever reach the stars, increasing evidence points to the existence of immense black holes that originate not in simple collapsed stars (in which case they are only a few miles in diameter), but in dark material weighing the same as millions of suns. The centers of many (perhaps all) galaxies harbor such supermassive objects. Like masked guests at a deadly costume ball, their brightly beautiful, falsely cheerful accretion disks would disguise their ebony hearts, while their huge size would dilute the alarming gravitational pull until too late.

The nearest black hole, however, can lurk no closer than a few hundred light-years, its unknown terrors too remote to have played any part in our genetic aversion to darkness. To understand that fear we need look no farther than our nearest neighbor, the moon.

A poor reflector of sunlight, the powdery slate-dark lunar surface is too miserable a mirror to offer usable light for animals or humans until it approaches the quarter phase. Even then it absorbs most sunlight, which is why the half-moon, amazingly, is only a tenth as bright as a full moon. For just five nights each month does the night sky glow with at least half the intensity of the full moon.

Real darkness is, however, rarely experienced in our country. The 10 percent annual increase in artificial lights that occurred from 1955 to 1985 has left less than 3 percent of the U.S. population in areas that are anything resembling "naturally" dark. Fear of crime has contributed to the sale of yard lights that remain lit all night long and that send almost as much light upward into the sky as downward, where it might accomplish its purpose. As a result, a milky synthetic skyglow is night's constant companion for most of us.

Analyzed spectroscopically, the glow is a combination plate made of sodium and mercury vapor, the components of street and parking-lot lights and yard lights. It is unseen anywhere else in the known universe.

When the level of darkness falls below about .03 lamberts, color vision disappears, because our retina's high-sensitivity but color-blind rod cells take over. This is why rummaging through a dark attic or basement is always a monochromatic adventure.

Yard lights typically throw light upward and sideways as much as downward to the driveway.

When lights have shields that reflect light downward, smaller bulbs do the job, saving electricity. There's also no glare to blind the homeowner or police when checking the yard. Extra bonuses: The night sky's glories are preserved, and the arrangement makes the house appear more attractive.

Actually, complete darkness is almost never available outdoors. Even during moonless nights far from cities in true wilderness, nocturnal vision relies not on starlight but on *airglow*—the natural radiance, like numerous miniature auroras, given off by our atmosphere (see page 123). On the darkest night, this still gives you enough light to see by, unless an overhanging forest canopy obstructs the sky.

As a child I wished I could experience the glories of the night sky from a vantage point in the clarity beyond our atmosphere. If the Milky Way so dramatically comes to life in a rural or desert setting, then what kind of fantastic presence must the stars exhibit from the moon or to an astronaut taking a space walk?

Happily, it turns out that we haven't been cheated after all. Our atmosphere is almost fully transparent to the band of visible light to which our eyes are sensitive. From pristine terrestrial sites such as isolated mountaintops or deserts, the stars are just a third of a magnitude dimmer than they'd appear from outer space—a difference only marginally detectable to the naked eye. And since beyond our atmosphere we'd always need a space helmet, the glass or plastic face mask of which reflects and absorbs so much light that stars would be dimmed by a factor of two or more, the bottom line is that space does not offer better stargazing after all! Nothing beats an unpolluted earthly site, especially when the air is dry.

Of course, while we usually imagine space to be the very paradigm of blackness, it's colorless; how could a vacuum be otherwise? Even if we teleported ourselves to some distant realm far from our sun, enough light would always arrive from the rest of the universe to disqualify it as perfect inkiness. So while we may fear the night, we are not fearing complete blackness: It really doesn't exist in nature outside of caves and such.

Our lone natural guardian against real problems is, of course, the sun. Without it, our planet offers no insulation from the icy horrors of space. That's a very different situation from the one we'd find on Jupiter or Saturn, which abundantly generate their own internal heat. We're spared a quick frozen death by the sun alone, a total dependency that is a bit unfortunate, since the latter has historically proven disconcertingly fickle.

Solar cycles do disintegrate, with serious terrestrial consequences. During the Maunder Minimum (see the chapter titled "Certainty Principle," page 95), there was widespread suffering. Nobody at the time linked the icy misery with the sun's activity, but abundant geologic evidence shows that the sun periodically varies its output for reasons unknown.

The very existence of terrestrial life proves that the sun is a fairly stable star. But the fragile sandbox of Earth is uninsured against solar upheavals, and even small sun variations translate into major misfortune.

Yet such sobering realizations seem to play no conscious role in our fears of darkness, as if our sun dependency were just a casual addiction removed from the night's experience. Instead, our modern focus is on control. We neither accept, surrender to, nor adapt to night; we overpower and synthetically defeat it. The vast, beautiful underground malls of Canadian cities, the dazzling neon-lit lure of Times Square—ours is an age in which we do more than merely cope with night; it has become important in and of itself, a time of leisure, of romance, of theater and dining. Shakespeare was perhaps prescient when, in *Romeo and Juliet*, he foresaw an era when "all the world will be in love with night, and pay no worship to the garish sun."

No, our fatal rendezvous with one of the night's (or day's) true dangers—collision with comet or giant meteor—is statistically many millennia in the future. The unpleasant prospect of such an encounter, and our fear of the night, is balanced by lack of short-term menace: For the lifetime of our society we are almost certainly safe.

Fortified with that statistical armor, we venture out. Even if our friends from the city, and our own genetic hardwiring, need time to catch up.

The Boundaries of Space and Logic

When I hear arguments over insoluble, abstract philosophical points, it creates love in my heart, love of humanity. For nowhere else in this vast left-handed biosphere of lunatic life do we find creatures who intensely care about questions that have no solution—and which, even if there were an answer, wouldn't affect them in any possible tangible way.

This tortuous business is exemplified by the square root of minus one. Here is something elegantly simple, a concept that anyone can visualize: What multiplied by itself results in the number negative one? The answer, of course, is that there is no such number. It doesn't exist. Yet it remains a mainstay of math, a genuine entity despite a pedigree from disembodied spirits.

So now consider basic cosmological questions: Where does the universe end? Does it have a boundary? If so, what lies beyond?

On one level, these questions seem as insoluble as minus one, and yet, unlike others that we'll get to shortly, they have been successfully tackled for nearly a century. In this particular case, the conclusion is that there is no single "edge" or boundary to the universe—there are *three*.

Einstein's general theory of relativity laid the foundation for our understanding of cosmic boundaries back in 1915 by describing space (actually space-time) as exhibiting characteristics such as curvature. If everything that weighs anything can warp space and time in its vicinity, then the sum of all the matter and energy that exists must surely exert a tremendous general warping on the whole shebang.

And it does. Einstein threw us all a curve by describing two possible types of warping to the universe at large. Which one proves to be correct depends on the overall mass and density of the cosmos. Because this latter figure appears to be extremely close to the critical level—called Omega—where the universe just barely stops expanding infinitely far in the future, we cannot yet know with certainty on which side of the fence we sit and therefore which type of curvature our universe exhibits.

If it contains several times more mass than presently detected, the universe has a *positive curve*. This is an inward curve, which means that an astronaut traveling near light-speed for billions of years would find him- or herself returning into familiar territory. Science writers often say that the astronaut would eventually come back to his or her starting point, but this is an oversimplification. Still, the concept is clear: No physical boundary would ever be encountered. There would be no decisive edge. Instead, by traveling in a seeming straight line, the voyager would eventually experience some of the same galaxies and stars that had already been passed aeons ago. It's like the time a traffic jam forced me off an expressway in the Bronx. After receiving several helpful but contradictory directions, I found myself passing the same bodega over and over.

Put succinctly, the universe is finite but unbounded. There is not an infinite amount of energy or mass. Nor does it contain unlimited volume or space. There is no barrier either. If that reminds you of Abbott and Costello's "Who's on First" confusion, stay with it awhile; the problem always lies in trying to grasp one additional dimension beyond the ones with which you're familiar.

A shadow creature whose existence was confined to just two dimensions would be incapable of grasping the meaning of solids; how, then, can we be faulted for being unable to picture four-dimensional space-time?

We can attempt to visualize a two- or three-dimensional analog to the universe's four-dimensional space-time by picturing a perfectly straight highway constructed along Earth's equator, using bridges where necessary to span the oceans. While we're familiar with Earth's near-spherical geometry, people are practically two-dimensional creatures compared with our planet's spherical vastness. Thus, a tenth-century motorist who believed Earth to be flat would not be surprised that the road would seem to stretch on indefinitely in a straight line. Set the cruise control and keep going directly ahead. It would come as a bewildering shock when previously seen landmarks were passed a second time. Only a person with advanced knowledge of Earth's three-dimensionality could explain what had happened.

The other type of curvature will hold true if there is a bit less mass and overall density than seems to be the case. This situation dictates a *negative* curve, shaped somewhat like a saddle. It is divergent and scatters the astronaut off in new directions. Yet, through it all, there *still* is no boundary.

Of course, the no-boundary idea makes us uncomfortable: In the highway-around-the-equator analogy, we can picture why the driver repeats the route. With the universe as a whole, we can start to imagine space curving and even agree that all paths need not lead to any sort of an "outside." But if we then visualize the universe as *seen from the outside,* it leads us to question what that outside zone contains, what surrounds our spherical universe. We're back at the square root of minus one.

Cosmologists would reply that such reasoning is invalid, since objects exist only within the space-time of the universe and simply

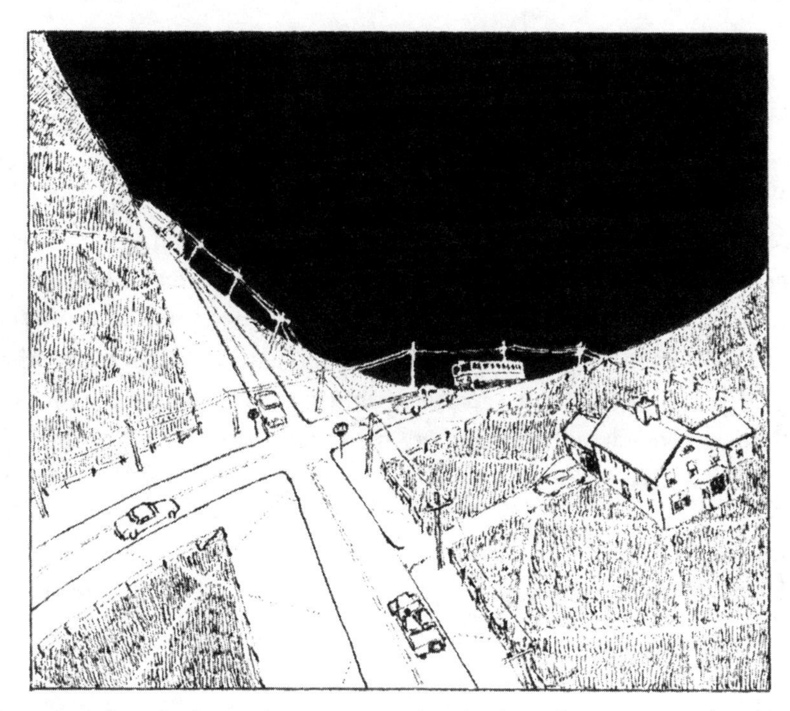

In negatively curved space, every seemingly straight path turns out to be a saddle-shaped (hyperbolic) curve.

do not exist apart from it. Space without matter to define it is literally an empty idea, as nongraspable as "*being* dead." (You can't be and not be, any more than you can walk and not walk simultaneously.) It *sounds* valid, but it's vacant. So there is no "outside" once you've defined the universe as "everything."

While such semantic wrangling very much belongs in the square-root-of-minus-one house of mirrors, other boundaries to the universe are far more logically clear-cut. Beyond the "curved space" boundary of a universe that is finite but unbounded, there is another, entirely different kind of frontier enclosing the cosmos. This occurs because of the red shift produced by the expanding space-time. Objects at increasing distances have their images dimmed and reddened by the Doppler shift, which causes their photons to lose energy. At extreme cosmological distances, where, we've learned, remote galaxies and quasars are receding at more than 90 percent of the speed of light, their light rays still manage to reach us—but they have paid a price.

Not only are all light-emitting and light-reflecting objects dimmed radically with increasing distance (light falls off inversely with the square of distance), but the very light itself has been weakened by the Doppler effect. At a particular distance, which probably corresponds to about 14 billion light-years (plus or minus 30 percent) objects would be receding at virtually the speed of light.

We then would not see them—but not because their light rays could not reach us. Remember, no matter how fast an object is

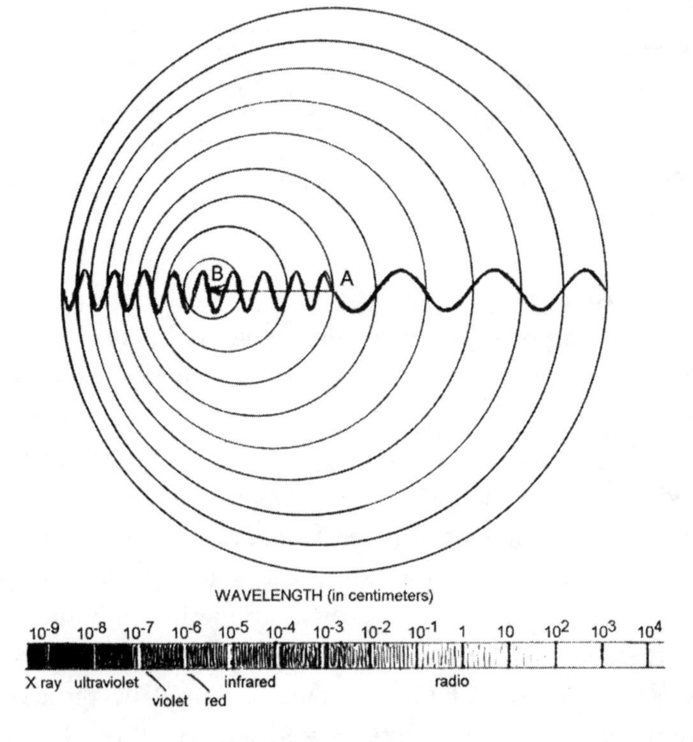

A galaxy, emitting light in all directions, travels from point A to point B at half the speed of light. As a result, the light directed backward displays an increased (reddened) wavelength, while forward-directed light is scrunched into shorter wavelengths and appears bluer.

Light appears dimmer as you move away from its source because it is spread over a larger area. Our technicians are confirming that the area covered is proportional to the square of the distance—producing a rapid fading of the light's intensity.

receding, its light will still arrive at light-speed. Instead, what happens is that so much of its energy is red-shifted into weaker and weaker parts of the spectrum that not enough remains for us to see. The visible light has been shunted into the feeble domain of the radio and finally the anemic microwave band, until the energy is too fragile for even the most sensitive of our detectors to perceive.

This, then, constitutes a real edge of the universe: The distance at which objects recede so close to light-speed that they are no longer detectable. Anything farther is forever invisible, beyond the horizon.

A third boundary to the universe is created by the finite age of the cosmos. There is a vast realm where objects cannot be seen because there just hasn't been time yet for their images to have arrived at Earth. The vast majority of galaxies were born long after the Big Bang. Our sun, if it is typical, was formed 4.6 billion years ago, some 10 billion years after the creation of the present universe.

So this third "edge" comes simply from the finite speed of light coupled with the youthfulness of the present universe. There is "darkness" beyond a certain distance because the universe is still so young that the news of most of it hasn't yet crossed the oceans of emptiness to our eyes and instruments.

These three separate limits are daunting. They do not seem amenable to even tremendous advances in technology. Unless we can fulfill the science-fiction dream of "folding" space, of jumping from here to there without traversing the space in between, nearly all of the universe will lie beyond one or more of the described horizons.

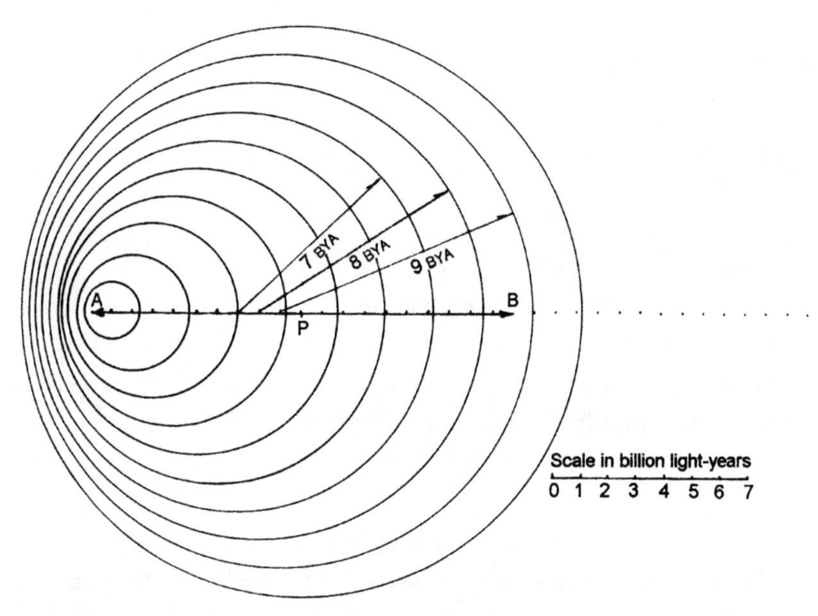

In an expanding universe, it takes a long time for images of distant galaxies to catch up to us.

Imagine that an external galaxy A, and our own galaxy B, left point P 10 billion years ago in opposite directions, traveling at three quarters of the speed of light. You might think that light from galaxy A would never reach us, since we are separating at 1.5 times the speed of light. But light always catches up with us eventually. Notice that the light emitted by A 9 billion years ago (labeled 9 BYA) has already passed us, so that we can presently view that galaxy as it was in the distant past. The light emitted 8 billion years ago will pass us in another 4 billion years. But the light emitted by A today will not catch us for another 60 billion years—a long time to wait before we learn the latest gossip.

And if most of the universe is forever hidden behind barriers, what then of other universes? How can we ever know whether or not they exist and what they're like? Are some things built by accident or design to lie beyond the human thinking system?

Why not? We can conceive of a metauniverse comprising trillions of lesser universes, each somewhat larger than everything that is now within view of our largest telescopes. If all that we see, the entire observable cosmos that started with a Big Bang, is just one item in a vastly huger universe, like cells within a body, we can have no inkling of what that metauniverse might be like for a surprisingly specific reason, one that goes beyond the fact that it's too remote to be viewed.

Just as quantum laws pertaining to the atomic and subatomic kingdom describe a wholly different reality from the macroscopic realm of the visible universe, and each enjoys its own rules and sets of logic, the metauniverse's properties and organization might have no parallel in the lesser dimensions.

Example: When sodium and chlorine, two virulent poisons, merge to produce the benign substance of salt, all their dramatic original qualities vanish without a trace. Sodium's explosive, incendiary reactiveness with water now becomes part of a substance that not only doesn't loathe water, but actually dissolves innocuously within it. Who could have predicted this? Similarly, the sum of ours and all other universes (if they exist) might constitute a new reality whose fundamental properties are wholly different from what we see around us, perhaps unimaginable in our most irrational dreams. In short, it's wrong to assume that we can extrapolate from our universe to conceive what the next level would hold in store.

As with the square root of minus one, there's something inherently unsatisfactory about all this: We start with speculation and finish up the same way, with no substantial answer. However, any inquiries that *always* produce unsatisfactory answers make me suspect that there's something lacking in the questioning process itself.

Instead of merely feeling frustrated by transparently weak explanations, I suspect that the human brain's logic system, with its linear thought, is the wrong tool for some of these operations.

The human brain evolved to handle things we can see and relate to, like Chinese food and flat tires. Is it any wonder its logic system fails with the subatomic—and may again with the macrocosmic? We've got a screwdriver when what's needed is a pliers, yet we arrogantly assume that every cosmic handyman's project we'll ever encounter requires a Phillips head. A little humility, please. As Niels Bohr said, "Every sentence I utter must be understood not as an affirmation but as a question." Perhaps patience will help, if we give our neural mass another half-billion years to evolve to adapt to the new cosmic complexities.

Limitations of our instruments are another matter and pen us behind fences built of different parameters—those of technological immaturity. Until the space age, many of these restrictions involved our atmosphere's opacity to most of the wavelengths of energy arriving from the rest of the cosmos. Now that space-borne instrumentation routinely monitors the sky, and the IUE (International Ultraviolet Explorer), HEAO (High Energy Astronomical Observatory), COBE (Cosmic Background Explorer), and Compton Gamma Ray Observatory, as well as many other satellites, have probed the hidden alleyways of the night, not one part of the spectrum remains unstudied.

Unfortunately, the very largest telescopes continue to be rooted to the Earth and will remain so for the foreseeable future. This means that not only are our most sensitive "eyes" blind to much of the universe, but the region we do see must be scanned through miles of soupy, churning air.

It is this degradation of image sharpness, rather than any technological limitation caused by the finite size of our optics, that has thus far restricted the discoveries contributed by the world's telescopes. The good news is that in this realm, major improvements are underway. Recently declassified from the military are *adaptive optics* techniques that let us sharpen out the blurriness of the air. With numerous tiny pistons located behind a telescope's mirror, the optics now can adjust to changes of temperature and rapidly compensate for image distortions produced by poor atmospheric conditions. Computer techniques augment this by rapidly calculating what the image should be like, were it not for the twinkling interference.

Computers have also become indispensable in the analysis, acquisition, storage, and manipulation of the data itself. Which means that just as the universe is becoming more bewilderingly complex, and our attempts to acquire better images and data require fantastically intricate processes, there have come to the rescue the computers and computer programs that allow us to pull it off!

Still, even here, in the theoretically do-able realm of technological challenges, daunting limits lie just around the corner. The Hubble Telescope has recently broken the magnitude-30 barrier by detecting galaxies and stars nearly a trillion times fainter than anything seen by the naked eye. That most of the universe appears much, much dimmer than this is discouraging, since we're nearing the limits of how much fainter a realm we can probe from Earth's surface.

The problem is airglow, our atmosphere's faint nighttime luminescence caused by the sun's ultraviolet light stimulating atmospheric atoms to emit a feeble flush. The night sky, from an unpolluted mountaintop or desert setting, is quite dark but not infinitely so. For extremely faint objects to be detected, they must be brighter than the night itself, and we have almost reached the point where this is not so. Then only space-borne instruments like the magnificent Hubble will help us continue to probe the cosmos.

So far we've looked at two regions—the seeming absolute limits of knowledge woven into the universe's very fabric and technological challenges that are merely a matter of time, money, and human cleverness. But the public wants more. People get fidgety with too much science alone; they want things to make aesthetic or spiritual sense. (Einstein, a product of an earlier, more religious generation, felt that tug and said, "Science without religion is lame, religion without science is blind.")

The popularity of books that attempt to wed astronomy or cosmology to grander questions attests to this innate longing for a Greater Viewpoint. It is in this theater that the play often has a short and unsatisfactory run.

The marriage of faith and science has always produced strange offspring, yet the courtship recurs time and again. One excellent illustrative example is the Christmas Star program annually shown at most planetariums. It always includes scientific explanations for the Star of Bethlehem: bright conjunctions of planets, a supernova, or a comet.

Planetarium directors are fully aware that none of these is a genuine "explanation," simply because nobody could have reached Bethlehem (or anywhere else) by following any heavenly body. As every sky watcher knows, celestial objects rise, arc across the heavens, and then set: Any follower would merely walk in a giant circle. The only exception lies in objects in the northern sky, which scarcely move. If the Bethlehem "star" sat near the north celestial pole, it would indeed hover, glued in place. Alas, the Magi weren't going north but southwest to get to Bethlehem, so we're back to the inescapable fact that no conceivable scientific explanation works.

This doesn't necessarily mean that the Magi didn't exist or that they didn't see a "star." If we're talking religious faith here, then a light could have appeared—flash—just for them. The silliness of the scientific attempts at justification should be obvious: If a supernova did just happen to go off at the right moment in just the right part of the sky so that three men would be guided to the Christ child, well, that kind of coincidence is indistinguishable from the miracle we were trying to explain away—we might as well skip the tortuous rational/scientific justification and leave the story just as it's told.

According to most scholars, the "star" was almost certainly a portent contrived astrologically, without any real analog in the actual night sky. But astrology quickly fell into disfavor with Christianity and science alike, so that this explanation isn't usually offered, not even in planetarium presentations. The end result, at planetariums, is an appealing but scientifically impossible accounting, known by all planetarium directors to be false yet nonetheless presented annually. All in a scientific setting. No comment.

Other efforts to wed religion or metaphysics with science have done nearly as poorly. *The Tao of Physics* was a best-seller that attempted to show that Eastern thought and modern particle physics are saying the same thing. Physicists almost universally rejected the book as fallacious oversimplification, but I have a deeper quarrel with it:

The nub of Eastern thought is that the underlying reality that transcends our dualistic perceptions is inherently ineffable. *Ineffable* means just that: You can't imagine it, you can't think it. You gotta *be there* to know what it's about. In other words, all the core, nondualistic teachings of Buddhism, Hinduism, and Sufi or Judeo-Christian mysticism are just pointers. Any cosmological or physical laws or concepts may espouse seemingly Eastern concepts (such as "All is one") but such phrases or ideas are definitely not the *experience* of transcendent perception.

We have a tendency to oversimplify, to assume that our thoughts will carry us to the edge of the lake, and from there it'll be an easier jump into the cool water. In reality, in any quest for higher consciousness, metaphysical practice involves fully seeing the futility of any and all concepts. Only then, when the mind fully grasps its own irrelevancy, does it drop away by itself. So long as an idea, whether Eastern or cosmological, carries a sense of validity to it—that's how long we will decidedly *not* drop the entire process of logical analysis and move on to whatever lies beyond.

So almost from the get-go the courtship is doomed.

We must leave them alone, faith and science, to go their separate and not always merry ways. Neither is appropriate for everyone's temperament; but for those with an interest in both, one offers nourishment for the mind, the other for a decidedly nonrational component. Their commonality runs about as deep . . . as the square root of minus one.

What's in a Name?

There's an obvious reason why English is considered the richest language: It has the most words. If you want to express yourself in Swahili, you can approximate what you mean, but you probably couldn't fine-tune it exactly. Here it's different. We can describe someone as dour, or we can say he's melancholy. Or dolorous, sad, saturnine, downcast, gloomy, dreary, morose, glum, woeful . . . it goes on and on. Foreigners may very well think that we need 200,000 words—forty times more than all the naked-eye stars in the night sky—just so we can say *bummer* thirty different ways.

Yet we lack some really helpful vocabulary. We're unnecessarily wordy when it comes to modern events, feelings, and situations simply because nobody's coined concise terms for a myriad of contemporary needs. There's no verb for the act of clicking a mailbox lid a few times to make sure the letter's gone down. No noun for that spot on a dog's belly that, when touched, will make his back leg twitch. There's no word for that cold, painful feeling in the head when eating ice cream too quickly.

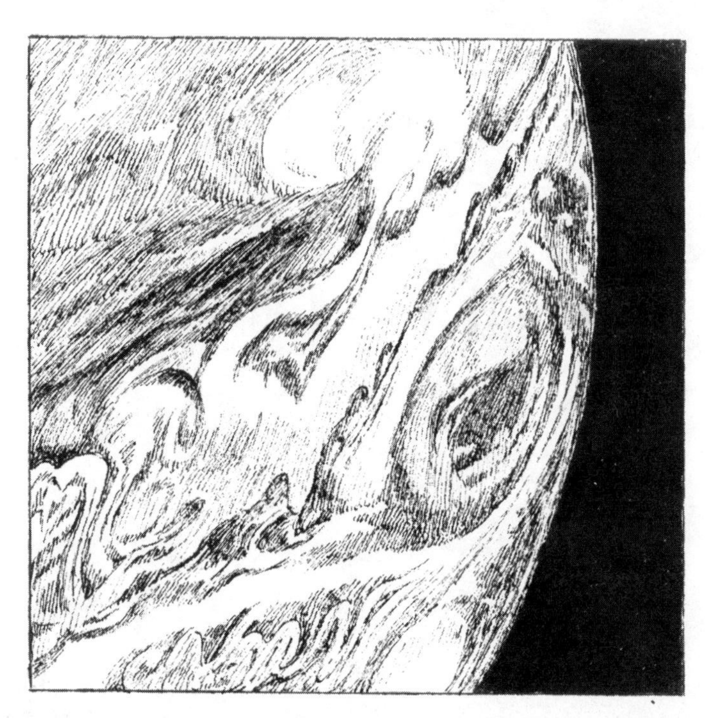

Jupiter's colossal storm, the Great Red Spot, seems to call for a name worthy at least of Poe's "Maelstrom."

Think of how you feel when seeing another car pull into a coveted parking spot just ahead of you. While there are fifteen different ways to say *sad,* there's only the single word *frustration* to cover the full spectrum of modern letdowns, each with its own rich palette of unpleasant nuance.

The same is true of space science. As the universe opens up to our instruments and to our minds, we remain saddled with an amazing lack of terms to describe the newly discovered phenomena and places. Worse, in many cases the existing names are misleading or insipid but remain curiously unchanged through the centuries.

With a breathtaking lack of imagination, what did astronomers name Jupiter's astonishing, enduring, larger-than-Earth cyclonic storm? Duh . . . the Red Spot. When the plucky *Voyager 2* spacecraft visited Neptune for the first time in August 1989 and revealed that giant aquamarine world to have its own semipermanent features, the most prominent was quickly and simplemindedly named the Dark Spot.

The rings of Saturn, beautifully bizarre 150,000-mile disks with an amazing, delicate thinness analogous to a sheet of paper the size of a city block, have no equal anywhere in the known universe. They can readily be seen through any good telescope as separate features, each with different widths, brightnesses, and degrees of transparency. Their inspirational names? A, B, and C.

The cluster of galaxies in which we live, a place where a trillion suns and probably an equal number of planets compete in loveliness with swirling scarlet nebulae of unspeakably intricate designs—this magnificent hometown group of thirty galaxies is called: the Local Group.

Apparently it wouldn't do any harm to have a few poets or naturalists join the naming committee of the International Astro-

The Great Dark Spot of Neptune whizzes westward at over 600 miles per hour.

nomical Union, which bestows terms and titles for nothing less than the contents of the cosmos.

Even the simplest one-word names for the universe's inventory tend to be archaic or lacking in creativity. *Galaxy* came from the old Latin term for the Milky Way, the Via Galactica, a reasonable description for the creamy, radiant glow that spreads across the summer sky. In fact, that ancient term *means* milk (notice *Galactica* contains *lactic*), so that, at face value, galaxies are just milky things, like cows, and we've essentially regarded the universe as one enormous dairyland.

Within our galaxy we find our *solar system*, a wooden phrase that does nothing to honor the sixty-six moons, eight planets, thousands of asteroids, and millions of comets and meteoroids (and also Pluto, whatever we end up deciding it really is; see page 156) that make up our celestial neighborhood.

As we look outward, we spot weird but common smoke rings surrounding many stars that have suffered relatively minor explosions, causing globes of colorful gas to blow outward like a child's bubble maker. These are all called *planetary nebulae,* though they have nothing whatsoever to do with planets.

And the numbing impoverishment of our cosmic nomenclature goes on. The center of our galaxy—the mysterious mansion in whose inner sanctum sits a black hole and around which our planet and sun and every star in the night sky revolve each 240 million years—is called the *nucleus*.

The precise genesis event that gave rise to the entire universe? Nothing more than a toddler's comment when playing a drum: *Big Bang*. The name for supergiant stars that alternately swell up like monstrous orange balloons and then collapse to perform the trick all over is as dull as expected: *pulsating variables*.

Half-million-mile surges of nuclear fire, whose tongues protrude from the sun like sinister entities (and whose violent subatomic drool licks our planet with myriad known and unknown consequences), are called *solar flares*. About as exciting as highway accident markers.

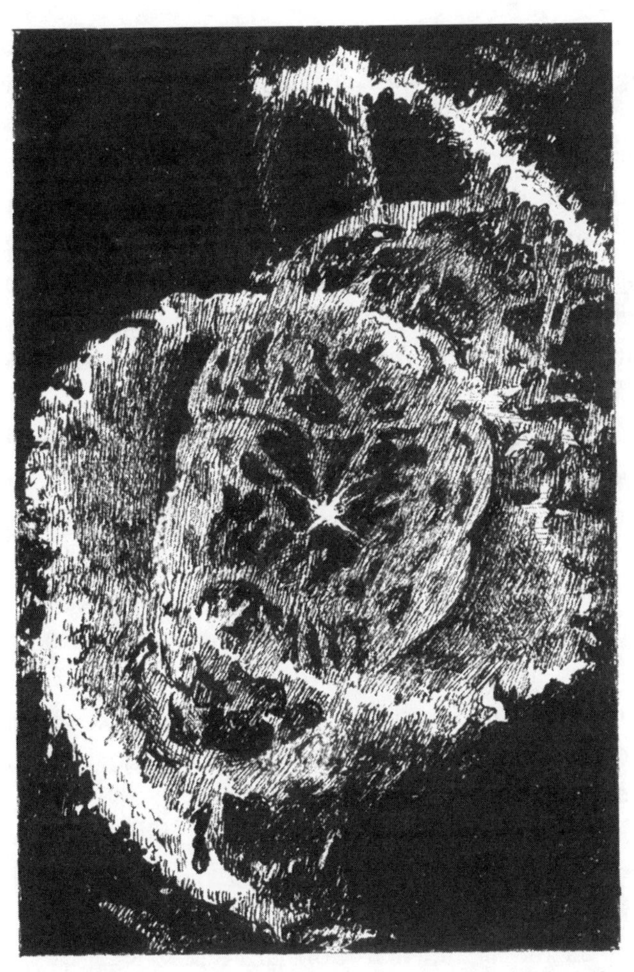

The Cat's Eye in Draco is a typical "planetary nebula" but has nothing in common with planets. This stellar outburst produced expanding bubbles and jets of gas that are violently overtaking the central star's earlier stellar wind.

We're accustomed to these terms, and in the deliberately dry realm of astrophysics, they serve just fine. Indeed, the few intriguing names that do exist are avoided in scientific papers. The great Andromeda Galaxy is merely referred to as M31. The brilliant pumpkin-color spring star Arcturus becomes Alpha Bootis. However awesome, everything is reduced to numbering and lettering schemes.

What's odd about it is that these nomenclature systems wound up being different for each type of celestial beast. Planets were named after Roman gods, moons for mythological figures. Asteroids were numbered in their order of discovery. Comets were named for their discoverer. Most stars have six- to nine-digit numbers indicating their position in the sky. Galaxies also have numbers, but these indicate where they're found in various catalogs. Galaxy clusters are named for their foreground constellation, larger and fainter clusters after the person who catalogued them, supernovas for the year in which they exploded. Variable stars have English letters sequenced for their order of discovery, followed by their constellation name.

You get the picture. It's more than mythical or poetic reference that's as absent as dark matter; there's also no rhyme or reason for the naming schemes, nor any sense of coherency.

Maybe that in itself is symbolic, even appropriate. It mirrors the universe itself, whose waterfalls and meteors follow no rational pattern, offer no predictability. In a mad cosmos around whose every turn awaits a wonderful strangeness, why should our nomenclature alone be standardized?

Still, a little playfulness or imagination, akin to that of physicists, who keep bestowing new subatomic particles with whimsical names like *strange quark,* might be a treat for those of us who spend time gazing at the (sigh) northern lights.

There have been attempts. The late Robert Burnham, who died penniless and forgotten after having written what is arguably the most epic and thorough treatment of the sky's wonders in his two-thousand-page opus, *Celestial Handbook,* tried here and there to offer noble names for various phenomena. He suggested, for instance, that the pillar of glowing gas called M16, which forms a vast structure somewhat resembling a person on a throne, be called the Star Queen Nebula. I immediately started using that term in my lectures and magazine columns, but so far no cigar.

A few years ago, *Sky and Telescope* magazine conducted a contest to rename the event that started the universe's expansion. After all, *Big Bang* had originally been coined pejoratively by a disbeliever in the theory, and while amusing in its understatement, it hardly seems worthy of cosmic genesis. After receiving thousands of entries, the judges decided that no suggestion was good enough; perhaps no word or phrase can convey something so inexpressibly vast and mysterious.

Other phenomena may simply be too esoteric to merit a new name. Violent, stormy pinwheels of billions of suns are certainly grand enough to deserve something better than the term *Seyfert galaxies*, and the weird upside-down rainbow that sometimes sits atop a solar or lunar halo ought to be called something more intriguing than an *upper tangent arc*—but such phenomena are too unknown or unobserved for most people to care.

At the other extreme, the most common celestial objects will never be renamed because their one- or two-syllable names—sun, moon, planet, star, comet, nova—are already so well established. Only the intermediate class of cosmic creature, which happens to comprise most of the sky's wonders, includes good candidates for complete makeovers.

The greatest hope lies in new discoveries. Astronomers who announce the latest findings might use their fifteen minutes of fame to suggest a suitable and memorable name instead of leaving it to the bureaucratic name-by-committee process. (Bureaucracy has been called "the epoxy that greases the wheels of progress.")

In my town, a restaurant named Dot's became Duey's and then Carolina's before its current incarnation as Wok and Roll. In astronomy, existing objects are much less likely to be renamed, no matter how contradictory we later find their appellation to be. Given the classic, immutable laws of intellectual inertia, we're likely to encounter alterations from the prosaic to the inspirational just once . . . in a blue moon.

Time and Time Again

Time is often called the fourth dimension. This usually throws people for a loop, because time in daily life bears no resemblance to the three spatial domains. Which, to review basic geometry, are:

Points, which have just a single dimension.

Lines, which are two-dimensional. (Except in string theory, which offers an exception to two-dimensional lines: Its threads of energy/particles are so thin, they're stretched-out points that do not quite constitute a second coordinate. The ratio of their negligible thickness to an atomic nucleus equals that of a proton to a large city.)

While a normal line has two dimensions, a sphere or cube has three. An actual sphere or cube requires four because it continues to endure. That it persists and perhaps even changes means that something else besides the spatial coordinates is part of its existence, and we call this time. But is time an idea or an actuality?

Time appears to be indispensable in just one area—thermodynamics, whose second law has no meaning at all without the passage of time. Thermodynamics' second law describes *entropy* (the process of going from greater to lesser structure, like the bottom of your clothes closet). Without time, entropy cannot happen or even make sense.

Consider a glass containing club soda and ice cubes. At first there is definite structure. Ice is separate from the liquid, and so are the bubbles, and the ice and liquid have different temperatures. But return later, and the ice has melted, the soda has gone flat, and the contents of the glass have merged into a structureless oneness. Barring evaporation, no further change will occur.

This evolution away from structure and activity toward sameness, randomness, and inertness is entropy. The process pervades the universe. According to nearly all physicists, it will prevail cosmologically in the long run. Today we see individual hot spots—stars—spewing heat and subatomic particles into their frigid environs. The organization that now exists is slowly dissolving, and this entropy, this overall loss of structure, is, on the largest scales, a one-way process.

Entropy does not make sense without a directionality of time, because it is a nonreversible mechanism. In fact, entropy *defines* the arrow of time. Without entropy, time need not exist at all.

And maybe it doesn't anyway, according to many physicists, who argue that time on the deepest levels of reality has no validity or purpose. Newton's laws, relativity, quantum mechanics—all function independently of time. These laws operate backward as easily as forward.

Metaphysicians, taking entirely different routes, have also questioned time's validity. The past, they say, is just an idea in a person's mind; it is no more than a collection of thoughts, each of which occurs in the present moment. The future is similarly nothing more than a mental construct, an anticipation. Thinking itself occurs strictly in the "now." So where is time? Does time exist on its own, apart from human concepts that are no more than conveniences for our formulas or for the description of motion and events?

Lacking both motivation and brilliance, I'll make no attempt to settle this ancient debate, but suffice to say again that thermodynamics provides the only strong argument for time's reality. The actual answer may be mind-bendingly more complex, because there may be many planes of physical reality. My own view is that time may operate on some levels but be nonexistent or irrelevant on others.

Physicists in the past two or three decades have also taken seriously the notion that the arrow of time can change direction. Even Stephen Hawking once believed that if and when the universe starts to contract, time would run backward. But he later changed his mind (as if to demonstrate the process). In any event, time running backward is not as screwy as it first seems.

We protest because we think that it means effect would precede cause, which never can make sense. A serious car accident would become a macabre affair in which injured people instantly heal without a blemish while their wrecked vehicle leaps back, uncrinkling and repairing itself seamlessly. This is not only ridiculous, it doesn't accomplish any purpose, such as, in this case, instruction in the evils of DWI.

The usual answer to this objection is that if time ran backward, our own mental processes would operate in reverse as well, so we'd never notice anything amiss.

Assuming that time is an actual state of existence, might it not follow that time travel should be valid as well? No. Very few theoreticians take seriously the possibility of time travel or of other temporal dimensions existing in parallel with ours. Aside from the violations of known physical law, there's this little detail: If time travel were *ever* possible, so that people could journey into the past, then—where are they? We've never been faced with tales of unexplained people arriving from the future.

Is this absolute proof that voyages backward through time will *never* be possible? Around a dinner table with some close friends, on a long, cold winter evening, it would be fun to try to debate this—if you have the time.

But we rarely do. Unlike other animals, humans seem aware,

even obsessed, with time and give it enormous importance despite its elusive nature. The same forty-five minutes that seem fleeting and ephemeral when spent with a fascinating companion drag on interminably in a traffic jam or for high schoolers serving detention. Even the passage of years or decades appears to vary its velocity.

Sometimes we marvel at being alive in this technological age but fail to have any real perspective on how transient is our era. After all, my grandmother, who died at age ninety-five, was born in the year 1893. She in turn, as a child, had spoken with people who had been alive in the late 1700's! In short, someone I knew conversed with people of the eighteenth century. Going the other way, I may live to be ninety-five, in which case I'll still be around in the year 2040, when I'll see and speak with children who will live in the twenty-second century and who are destined to know people who will live in the twenty-third! So in just my one lifetime I will have known people whose lives spanned the nineteenth to the twenty-second centuries and who will directly interact with humans from the eighteenth to the twenty-third centuries. Generations whiz past like the shuffling of cards. If we are fifty years old, then the way things were when we were children seems somewhat different from today but not radically so. The same half-century span, when considered historically, seems enormous: What a stretch from 1899 (before the first airplanes) to 1949 (after the atomic bomb)!

Regardless of whether time exists on all levels, its passage certainly varies in perception and definitely alters in actuality. We point telescopes to places where we can *see* a more lethargic unfolding of time and also observe places as they existed billions of years ago. Time's makeup seems as strange and elusive as that of sausages.

Let's try to clarify one common alteration in the passage of time with a simple thought experiment. Pretend you're blasting off from Earth, looking out your rocket's rear-facing window, telescopically observing the people near the launchpad, who are applauding the successful liftoff. Each moment you are farther from them, so each moment their images have a longer distance to travel to your eyes and are therefore delayed, arriving significantly later than the last "frame" of the movie. Result: Everything appears

A single lifetime's flight evolution, from the 1910 Wright biplane going 34 miles per hour, to the 1976 Lockheed Blackbird at 2,193 miles per hour, to the 1982 space shuttle at 25,000 miles per hour. But we're still only traveling 1/26,000 the speed of light.

in slow motion, their applause dishearteningly lukewarm. Still with us?

You see now why nothing speeding away from us can fail to appear in slow motion. And since more than 99.9 percent of the universe's contents *are* receding, we're peering at the heavens in a dreamy kind of mandatory time-lapse photography; the unfolding of nearly all cosmic events takes place in a false time frame.

This was exactly how the speed of light was discovered, by a Norwegian named Ole Roemer more than two centuries ago. He noticed that the moons of Jupiter slowed down for half the year and, realizing that Earth was then moving away from them in our orbit around the sun, was able to calculate light-speed to within 25 percent of its true value. Conversely, those satellites would seem to speed up for the other six months, just as inhabitants of an alien world would go about their business at an accelerated fast-forward, Charlie-Chaplin pace as viewed by approaching astronauts.

Superimposed on these illusory distortions that are nonetheless inescapable is the actual slowdown of time at high speeds or in stronger gravitational fields. This is not merely something we can shrug off with facile rationalizations, like an errant spouse's late homecoming. This zooms to the far end of peculiar.

This *time dilation* effect is minor until one nears the speed of light; then it becomes awesome. At 98 percent of light-speed, time travels at half its normal speed. At 99 percent, it goes just one seventh as fast. And we know this is true; it's real, not hypothetical. For example, when air molecules high in our atmosphere get clobbered by cosmic rays, they smash apart like the breaking of a stack of billiards, their innards spewing earthward at nearly the speed of light. Some of these subatomic bullets pierce our bodies, where they can strike genetic material and even cause illness.

But they oughtn't to be able to reach us and do such villainy. This atomic material is so short-lived, it normally decays harmlessly in a few millionths of a second—not long enough to travel all the way to Earth's surface. It manages to reach us only because its time has been slowed by its fast speed: Its extended fantasy world of false time allows it to enter our bodies. So relativistic effects are far from hypothetical; they have often brought poisoned offerings of death and disease.

Travel in a rocket at 99 percent the speed of light and you'll enjoy the consequential sevenfold time dilation. From your perspective nothing has changed; you have aged a decade in ten years' worth of travel. But upon returning to Earth you'd find that seventy years have passed and none of your old friends are still alive to greet you.

Then the truth rather than the theory will have hit home: Ten years can really pass for you and the rest of the crew, while *at the same time* seven decades elapse back on Earth. Abstract arguments then fail. Here a human lifetime has passed, while there it's been only a decade.

You might try complaining that time is supposed to have no preferred state—how, then, can nature determine who should age faster or more slowly? In a universe without privileged positions, couldn't you claim to have been stationary while the Earth moved away and then came back? Why shouldn't Earth's inhabitants be the ones who aged more slowly? Physics provides the answer.

You were the one who lived longer, therefore the answer must lie with you. And it does: It was you who felt the acceleration and deceleration forces of the trip, so you cannot deny that it was you and not Earth that made the voyage. Any paradox is nipped in the bud; the one who traveled also knows that it is he or she who should experience the slowing of time.

Einstein taught us not only that time mutates, performing its own unique rite of passage by varying its rate of passage, but that distance contracts as well—a totally unexpected phenomenon. Someone zipping down the freeway at 99.999999999 percent of light-speed experiences a dilation effect of 22,360. Her watch ticks off 1 year, while, simultaneously, a severe traffic jam lasting 223 centuries elapses for everyone else. In addition, all distances in front of the speeder shrink by this same margin. If we're contemplating a more enchanting destination than downtown Los Angeles—say the galactic nucleus some 26,000 light-years away—that intriguing storyland is now reached in little more than a single year: To the speeder it's situated a mere 1 light-year ahead. The round trip involves an investment of just two years, though a disheartening 520 centuries elapse simultaneously back home.

Another strange outcome of high-speed travel is that near the speed of light, *everything* in the universe would seem to lie directly ahead! This bizarre wrinkle comes from the entirely separate effect of *aberration*.

When we drive through a snowstorm, the flakes appear to come from in front of us, while the rear window hardly gets hit at all. The same thing happens with light. Our planet's 18-mile-per-second motion around the sun causes stars to shift position by several seconds of arc from their actual locations. As we increase our velocity, this effect grows ever more dramatic, until at just below light-speed, the entire contents of the cosmos appear to hover in a single blindingly bright ball, dead ahead. Looking out any other window, you would see nothing but a strange, absolute blackness.

Science fiction loves to show future astronauts traveling at light-speed, but in reality such a velocity can never be reached because of something that has no escape clause, no appeal: *mass increase*. At 99 percent of light-speed, a rocket weighs seven times more than at liftoff, and this heaviness increases more dramatically with continued acceleration. At a whisker below the speed of light, a spacecraft outweighs the entire galaxy, and then the universe itself! Any slight increase of velocity, even boosting speed by an additional snail's pace, would demand more power from the spacecraft's engines than has ever existed.

Obviously, hyper-light-speed travel requires more than mere technological cleverness. You'd have to travel into another universe that has different natural laws.

But none of this is even necessary. For nobody *needs* to break

By the time your spaceship achieves 94 percent of light-speed, the view from the bridge is pretty dull.

the light barrier. Merely approaching it would achieve all that one could want—to travel anywhere in the cosmos in a single lifetime. But it would be a lonely voyage. Earth and all its inhabitants would meanwhile experience their customary prosaic passage of time. A trip to the nearest galaxy could theoretically be accomplished in just a few years. But there'd be no escaping the fact that 4½ million years would have simultaneously passed back home. Ancient texts that might contain legends of your departure would have long crumbled. Evolution would have altered the very species that was once your kin. You'd return as a primitive, an animal, an evolutionary throwback.

Not only would there not be anybody resembling you or knowing your language; to the new Earthlings, your presence would be as a Neanderthal who suddenly appeared on today's White House lawn. You'd be lucky if you escaped incarceration in a zoo and if they had something digestible for you at feeding time.

Perhaps eventually there will be craft that *can* travel forward in time, but we must accept the penalties for fooling with this phantom: consequences that are strange, severe, and as inevitable as the surrender of our beliefs. All these effects deal with relativity, the comparison of your time systems with someone else's. But the temporal enigmas go deeper still: Within any given frame, the reality of time itself is uncertain on the deepest levels. Underlying these bizarre relative mutations lurks the prospect, even the physical likelihood, that time simply does not exist at all.

In all probability, then, in the heart of space, there is no time . . . to lose.

What Else Is New?

New! Improved!" shouts the pitchman from the TV. The actor doing the commercial may not realize that he himself is new and improved from humans just a few millennia ago. The commercial hits home—an effective theme because it resonates with a very fundamental part of nature itself, from Earth to the farthest galaxies: New is unavoidable.

Recently, I reached for a reference book on a high shelf and instead dislodged a heavy piece of petrified wood that served as a bookend, which smashed into my shoulder on its way to the floor.

Despite the pain, I had to laugh. Being bruised by a 50-million-year-old tree fragment was an experience that intrigued me as much as the guilty notion of having received some deferred karma, punishment for filching it from the petrified forest thirty years earlier. But what was funny? Not the pain (no masochist, I) but the novelty. Don't we all get off on truly new occurrences?

A quest for originality drives nearly everything we do, in the sciences as well as in everyday life. *New!* goes far beyond that advertising cliché. Evolution is forever ensuring that *everything* is new and improved. And while we're accustomed to technology speedily reworking our daily lives, we give little thought to the possibility that the same procedure may be operating throughout the cosmos. Orion and Andromeda seem monuments to inertia, but behind their frozen jeweled faces the heavens keep cooking up new recipes, and this while under restraint to use the same ninety-two elements over and over.

Unfortunately, when it comes to the stars, "What's new?" can never be answered, since the latest events are images that, though traveling at light-speed, take millennia or aeons to arrive at our telescopes. Galaxy light delivers information about objects or events that is no longer valid, dramas that ended their Broadway run before the first dinosaurs were born.

It's always fun to imagine the technologically advanced civilization watching us with some supertelescope. If they could pinpoint individual humans and were observing this moment from a planet orbiting one of the stars in Orion's Belt, they'd be watching people's lives during the Roman Empire. They'd conclude that here is a species with no ability or inclination toward space travel, nor knowledge that stars are distant suns that do not circle around Earth. Such a violent, warring, parochial society as ours might hold the interest of their specialists in primitive life-forms, but we'd surely not be treated as peers worthy of a dinner call.

And, of course, it works the other way as well. If we finally do succeed with the SETI (Search for Extraterrestrial Intelligence) project's desire to find beyond-Earth civilizations, the signals' delay-time will always mean that we are "contacting" the past with unknown consequence. We'll have no way to know whether the current, new reality of these creatures is truly reflected in their transmissions of bygone eras. The paranoid among us might point out that even if we believed benign assurances of peaceful intent, their present rulers might right now be dispatching warships.

So "new" or "present" or "now," as Einstein observed, always involves distance as well as time.

New seems to attract new. The young person itches for activity and change, and would sooner give up the security of home and job than stay past when it bores. Conversely, one symptom of a person getting old in spirit is a preference for routine: Familiar is safer.

My first wife's father was routine's embodiment. He was an elderly French banker who'd lived his entire lifetime in the family's stone-and-marble house in Cannes. We once spent the summer with her parents, and they took us to a restaurant in the city's main nightlife strip, the famous La Croisette. Looking around with some interest, he remarked that he'd never been there before at night. I stopped in my tracks. "You've lived in this city since birth, you're now sixty-five, and you've never been downtown at night?" Yet it was true: He was such a creature of habit, it was simply outside his routine.

Except for the timeless ticking of the pendulum clocks, silence prevailed within the stone corridors of the house in which my wife had been raised. Meals never deviated by a minute: Lunch was served at noon, dinner at seven on the dot. I take pride in having preserved my sanity through the long weeks of that gothic visit.

My in-laws belonged to that class of people, mostly elderly, who vacation in the same place every year, seldom make new friends, and speak to strangers only on rare occasions and then only if absolutely necessary.

Young people, by contrast, are much more inclined to be curious and outgoing. Jobs and friends may change frequently; the world seems enormously full of new and exciting places. Why go back? Been there, done that.

Children and young animals exemplify this love of novelty. From the toddler to the adolescent, everything gets explored. First the room, then the basement and attic, then the neighborhood by bicycle, and finally other cities and perhaps the world. No wonder we have a built-in itch to visit the planets!

Of course, with quirky humans involved, the matter doesn't stop there. As if to throw a bewildering monkey wrench into this intuitive need to explore, society attaches a tight, conservatively antithetical leash. From early childhood, the same parents who encourage "original thinking" also insist that the child's actions stay within a narrow range of sanctioned performance. We impose our own values, most involving a good deal of rigidity. At home and at school, the goal seems not unfettered exploration or imagination, but a permanent state of tedium.

If I were setting up a school, it would use a Frank Lloyd Wright approach. Just as that visionary designed houses that harmonize with their environment, why not have education exploit the youthful need to *actively* explore the unknown? Of course, some educators succeed at exactly this approach, which makes such sense that *hands on* and *whole language* have become educational buzzwords; unfortunately, only the smallest fraction of school curricula effectively put it into practice.

So newness is prized and almost-new is a yawner. You're either first or you're nothing. History has scarcely bothered recording the people of the second team to climb Mount Everest, or the second person to sail the Strait of Magellan, or the second person to run the four-minute mile. Such feats were just as difficult for that second individual as they were for the first. And if two thousand years passed without anyone running a mile in four minutes, why shouldn't someone who accomplishes this a month after someone else be accorded equal or at least near-equal respect?

Of course it doesn't work that way, because breaking barriers is everything. Whenever the chairperson at a formal meeting asks for someone to "second the motion," hands go up but without enthusiasm. There's no distinction for those seconding something proposed by someone else.

In the United States, the presidency has become a post akin to royal privilege, while the also-ran vice president has traditionally played a somewhat empty and historically maligned role. The mock seriousness and pomp accorded the veep reminds me of the fable about the emperor's new clothes.

New is also a buzzword in the sciences: Nobody wants to be the second there, either. Nowhere is the heartbreak of being second more obvious than in the detection of comets. This is the final democratic area of the sciences, where honors and status don't even get you on the subway. Whether Harvard professor or convicted felon, anyone who first discovers a comet will have it bear his name. If the comet is one of the once-per-fifteen-year visitors great enough to emblazon an awesome tail across the heavens, its discoverer will become world-famous overnight and his name known ever after.

This has been true for centuries. In eighteenth-century France, comet hunter Charles Messier stood on his Paris rooftop night after night hoping to be the first to find the faint, tailless, smudgy ball that is the initial sign of an incoming comet. He was well aware that if, with the passage of some weeks or months, the comet should become spectacular, he was likely to receive patronage or even royal appointments. It was the quickest route toward winning the *jeune fille* and the million francs.

Trouble was, Messier kept finding telescopic smudgy spots all over the night sky, teasers that got him all worked up for nothing. Always, after a few nights observing one or the other, he'd sadly realize that these potential fortune-makers were glued in place. They didn't move, and they weren't comets. He didn't know *what* they were, and it didn't matter: He was getting tired of being fooled. So he made up a list of the 103 worst annoyances, the fuzzballs to avoid.

He never did discover a great comet, but Messier's nuisance list won him posthumous fame that lingers to this day. (How odd life is!) Better telescopes eventually revealed each of those smudgy balls to be remarkable objects. The first item on his list, Messier number one—called M1 since the early nineteenth century—proved to be the Crab Nebula, a twisted filamentary remnant of an exploded supernova. M31 proved to be the nearest major galaxy, Andromeda. M42 is the Great Nebula in Orion. In 1996, researchers reported that a colossal spherical galaxy in Virgo named M87 had at its heart a supermassive black hole weighing the same as 3 billion suns! It made the front pages all over the world, though few realized what that letter "M" signified. But there was Charles Messier's legacy. Deferred, but greater than if he had found a comet. Though he was behind a thousand others in the comet-discovery department, he was inadvertently number one in cataloging dozens of the most amazing things in the universe—even if to him they were mere annoyances.

Others were not so lucky. The astronomical world is filled with those who came in "place" or "show"—people who spent hundreds of sleep-deprived hours searching for comets, only to independently find one a few hours after someone else. Can you imagine the frustration of watching it brighten, and then listening to endless interviews with the "discoverer," when in most cases the

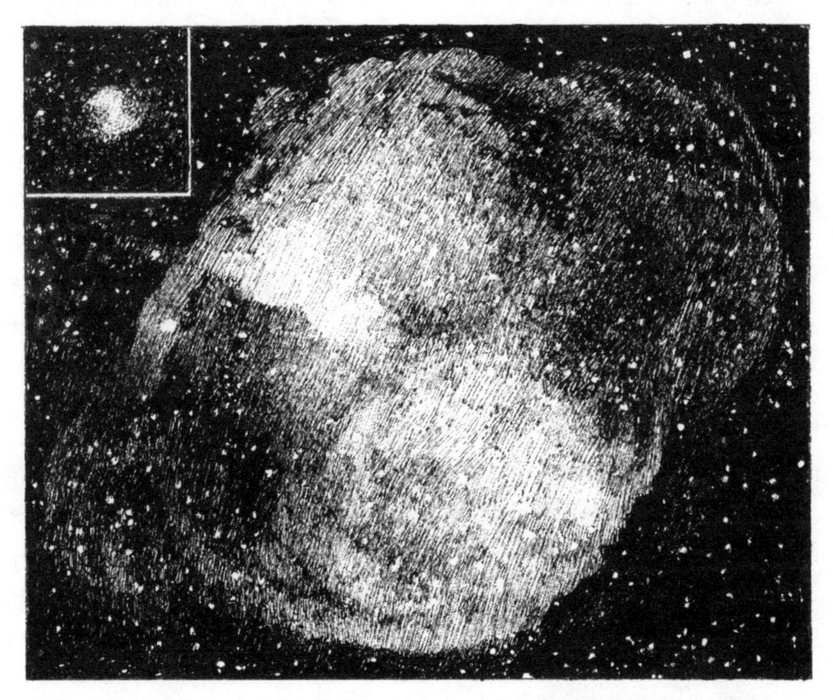

Charles Messier first saw this curious object on July 12, 1764. It looked like a comet's head but didn't move, so he listed it in 1771 as number twenty-seven in his catalog of noncomet annoyances. Commonly called the Dumbbell Nebula, it is recognized today as a gaseous shell three light-years across. Ejected by a dying star, it is expanding at 19 miles per second.

glory of first discovery would have been theirs but for the inopportune turning of the Earth, which happened to place some Japanese or European observer in the better position to face the brightening intruder a few hours sooner?

In March 1997, when Comet Hale-Bopp blazed its now-famous trail across the heavens while the world watched, I was privileged to join its co-discoverer, Tom Bopp, and renowned comet expert Fred Whipple as lecturing astronomers aboard a "comet cruise" in the Caribbean. Bopp and I spent the week showing guests his prize, and I loved his awed and bemused pride at his namesake's nightly performance. He'd been in the jewelry business, an amateur with a bad back, average lifestyle, and adoring wife, when he stumbled upon the largest comet ever known. Its unprecedented brightening, which allowed the detection of complex organic compounds never before seen in those pristine primordial balls of ice, propelled this modest, soft-spoken man onto the lecture circuit and bestowed invitations from leading observatories and even to attend a NASA space launch. The odds against it happening to him were far greater than those of his winning the lottery.

Discovering exploding stars is another arena in which ordinary stargazers often come in first. Distant galaxies are home to billions of suns, whose light melts into a homogeneous nebulous glow, but should a single one of those stars explode, the pinpoint is immediately obvious. Often I'd gaze curiously at such a single star, lying like a jewel against some galaxy's structure, and I'd turn from the telescope's eyepiece to check it out in a detailed atlas. Always it proved to be a foreground star in our own Milky Way, nothing more.

Then, in 1994, it suddenly happened: A new, bright, foreground star appeared in one of the best-known sky cities, the widely observed Whirlpool Galaxy. Though one of my favorite targets, it hadn't been on my list that night, just when the supergiant sun chose to blow itself into kingdom come, lighting up its galactic neighborhood like an explosion at a fireworks factory. Like a million other observers, I gazed at the supernova with awe, not destined to be the one to see it first.

Nova means new, so it's ironic that such novas and supernovas invariably arise in the final stages of a star's life. Like beneficiaries of some Madison Avenue marketing ploy, they eke out novelty from an object that's been sitting on the celestial shelf, literally gathering dust, for quite some time.

It's not easy to uncover truly new phenomena. Maybe that's why, in popular astronomical articles in magazines and newspapers, the word *new* is sometimes used over and over for the same kind of discovery. For years, headlines proclaimed, FIRST NEW PLANET FOUND AROUND ANOTHER STAR. As the initial announcement gave way to dispute and controversy, we'd become less sure about the initial data. Maybe it was a planet, maybe it wasn't. So when another researcher made the same proclamation a year or so later about a different discovery, it was treated as new and true all over again, until it, too, was disputed. These days, FARTHEST GALAXY FOUND is another recurring favorite.

Sometimes being first is almost effortless. In 1978, what was probably the easiest Nobel Prize in history went to Robert Wilson and Arno Penzias, two researchers at Bell Labs in Holmdel, New Jersey. They were merely calibrating the background noise of the sky when they found a hiss they couldn't get rid of. Through a series of fortuitous conversations with others, they eventually realized that they'd detected the background noise of the Big Bang explosion that had created the cosmos!

At first, they'd surmised that pigeons nesting in the radio telescope might be responsible. But shipping them away and cleaning off the droppings didn't make the sound vanish. From cleaners of bird doo-doo, they found themselves elevated to the status of the men who uncovered the Birth of Everything.

There was no discussion. There couldn't be any second place. Who could follow that act? You'd have to be the first to photograph heaven or hell.

Even a man as renowned as Galileo got into trouble over an authorship dispute. He claimed to be the first to discover sunspots, those speckles that actually had been noted in Oriental records for many preceding centuries. After all, the larger ones appear to the naked eye when the sun is red and subdued at sunrise or sunset. But nobody had ever mentioned it in Europe, at least not in writing. Suddenly, however, when the telescope was first turned skyward in 1609, Galileo and another Italian observer nearly simultaneously claimed first discovery and suffered decades of bitterness over the matter. According to an eyewitness, the competitor's face matched the sunset's crimson; he became apoplectic when he first saw Galileo's claims in print and later used Church connections to get Galileo into serious hot water.

The fight to be the first with what's new (with precedents reaching back to our animal ancestors, who butted horns for mating rights to the prettiest female, and progressing to the only marginally more civil enmity and rivalries of competing research teams today) has obviously been programmed into our brains. Sadly, the new is ephemeral by definition; there is no peace to be found in its obsession. But perhaps the fixation is not ours alone. If we are indeed constructed by nature to mirror larger themes—such as the continual creation of new stars from the deaths of old ones and a galaxy-wide evolution that constantly cooks up a fresh menu of new planetary designs—then our human craving for novelty will not go away, ever.

The obsession with newness is just too old.

The Most Astounding Discovery Ever

There's no topic—black holes and supernovas included—that makes us sit up and take notice like the discovery of a new planet. Maybe that's because we live on a planet, not a moon or a quasar, so that such an announcement literally hits close to home. But maybe our intrigue also owes its origin to a sort of collective genetic recollection of the most astonishing scientific announcement of *all time*—the finding of Uranus in 1781.

Whoa. Uranus? Astounding?

Today, that giant aquamarine dollop of hydrogen inspires as much awe as a cheeseburger. On our lists of the most exciting places to explore, Uranus sits somewhere near Paramus, New Jersey.

But imagine yourself a learned resident of the late eighteenth century. The age of science and reason is well advanced. New lands are being probed and colonized. The industrial revolution has transformed everyone's life. In an era of new discoveries, what could excite the jaded sophisticate?

Here's what: the realization that some simple, basic aspect of reality had now suddenly changed forever.

The existence of five and only five planets had been taken for granted for a hundred thousand years, since the dawn of sky-watching by semihumans dressed only in furry underwear. It was a certainty. Nobody was looking for any others, any more than scientists today are searching for signs that we are all actually androids or that the sun is hollow and filled with broccoli.

Some things just seem totally secure from reinterpretation, absolutely safe from further revision. Just as today we would not expect science ever to alter its opinion about the basic function of the human heart or announce that clouds are nutritionally valuable, citizens of the eighteenth century were equally certain that our solar system, by design or axiom, contained one Earth, one sun, and five planets. Period.

The utterly unexpected discovery of a giant new planet was more than a revelation or even a bombshell. It pulled the rug of intellectual security from under everyone's feet. If there were more planets, then no knowledge was secure. It was the most humbling and frightening experience imaginable for the world's educated citizenry.

Perhaps a shock that goes so deep can affect us all the way to our central nervous systems and somehow collectively linger like a traumatic childhood event. Of course, back in 1781, the finding of Uranus first set in motion other, more immediate endeavors. After the initial astonishment settled a bit, self-flagellation took hold: Why hadn't anyone seen this before? Uranus is dimly but clearly visible to the naked eye. It changes position. How had it managed to escape the notice of all the supposedly keen-eyed Arab desert-dwellers who had named the stars in the first place? Where were the Chinese observers who'd chronicled the supernovas when the Europeans were sleeping, during the Dark Ages? And what about the first 180 years of telescopic observation: Uranus is positively brilliant through any good telescope—why had it taken so long for

Uranus and its thin rings have their axes of rotation tilted to 8 degrees below the planet's orbital plane. Summer at the north pole was expected to create large atmospheric temperature differences as the sun bore down unrelentingly for twenty-one years, but the temperature remains surprisingly even from sunlit pole to shaded equator. The vast heat generated in the planet's interior may explain the uniformity.

someone to notice its motion and its green unstarlike disk? How could this have happened?

The first collective impulse was to name the new world Herschel after its discoverer, an apt reward that nonetheless drew a storm of criticism, and not because of its odd unplanet-sounding resonance. If Herschel sounds alien to us now, rest assured it's no worse than Uranus, the most mispronounced celestial body after

Betelgeuse (few correctly say YUR-an-us or BET'l-juice). But the tradition of planets being labeled after Roman gods was too ancient to be ignored, so Uranus eventually prevailed and everyone got used to it, more or less.

The discovery also opened floodgates of near-frantic searching for yet more planets, a celestial gold rush that didn't pan out for an entire human lifetime. When the dust settled two centuries later (that is, now), both the arduous hunt for various planets X and the fruits of those early intensive labors proved anticlimactic: Turns out there's just one other planet in our solar system, after all—Neptune, found in 1846. Pluto, uncovered in 1930, was long considered yet another planet, but most astronomers now agree that it's much too small and perhaps merely represents the largest example of a new, tiny class of cometlike ice chunks that inhabit the outer solar system. Its final classification remains to be determined.

It's not easy to find many other examples of truly unsettling, foundation-rattling knowledge. Laypeople and astronomers alike expected to find eventual evidence of new *families* of planets be-

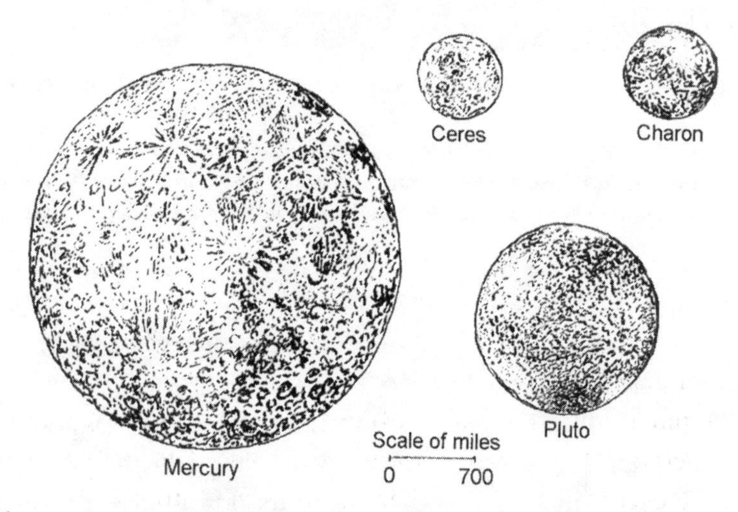

Is Pluto a planet? Compared with tiny Mercury, once thought to be the smallest planet, Pluto is a shrimp, even with its satellite, Charon, thrown in. Seven moons of the solar system are larger than Pluto, which is not too much larger than the asteroid Ceres.

yond our solar system, so the ones announced in 1995 and 1996 (which cannot be imaged, only inferred from their gravitational influence on their parent stars) were far from astonishing.* And, in fact, it's hard to think of anything else, in any other celestial or scientific arena, that astounded the general public.

Quasars were amazingly violent and distant objects of unknown characteristics but turned out to be mere explosive cores of ordinary young galaxies.

Penicillin was a wonderful discovery, but bread molds had been used in folk medicine in various countries for centuries. New elements (helium wasn't found until the late nineteenth century) were attention-grabbers but proved to be mostly inert substances that didn't affect the lives (or opinions) of the average citizen.

The airplane's invention was amazing in a way, yet people had already been airborne in balloons, so it was an evolutionary rather than a revolutionary leap. I'm not trying to be cavalier about great technological advances, but few if any announcements have made the world gasp.

The laser? Sure, it's everywhere now, as it reads supermarket bar codes and delivers music on our compact discs, but it didn't initially stun society or make the average citizen expect his life to become much different. The same can be said of the first automobile (initially regarded as a toy and not particularly practical) and almost everything else in the science world.

Yes, there was almost universal astonishment at the first Earth satellite, orbited by the Soviet Union in 1957, but it wasn't a bolt from the blue; educated people had long known that the United States had been trying to accomplish this. The only real surprise was that another country had beaten us to the punch.

Despite (or maybe because of) the fact that we live in this whirlwind technological era, it's paradoxically hard to recall a single scientific event or discovery that has shaken us to our underwear

*In 1998, a giant "runaway" planet escaping from its binary sun system in the constellation Taurus became the first planet outside our solar system to be photographed—by the Hubble Space Telescope.

in a single dramatic moment—not the computer, not the advent of television, not even the rocket. The things that most reshaped our daily lives sneaked up on us; they didn't leap out and yell "Boo!"

It's true that scientifically savvy laypeople were indeed left speechless when an entirely new state of matter—the Bose-Einstein condensate—was announced in 1995. But since it can only exist at temperatures within a fraction of a degree of absolute zero (−459.67 degrees Fahrenheit), it will not likely be seen in practical applications for a long while yet.

And it's also true that people into cosmology were astounded by the unexpected finding, in the 1980's, that the universe is structured like a sponge, with vast empty spaces surrounded by curving sheets of galaxy clusters.

Yet such truly unanticipated findings hardly touched the consciousness of the average citizen. No, to duplicate the discovery of Uranus, we'd have to stumble upon a finding so basic that everyone could grasp its meaning. It would have to displace cherished, long-held assumptions; it would have to instantly reveal that we had been dead wrong about some simple concept of everyday reality.

Thus, amazingly, while science and technology advance with exponential rapidity, our capacity for astonishment shrinks. Commercial interests now drive experimentation, and what we are likely to buy or to want is far more important to researchers than things that can have no possible pecuniary value. We want science to pull at us like a magnet, to intrigue rather than puzzle. We are in an era in which we are wooed more than wowed.

Uranus, whose sudden and totally unexpected appearance in our collective consciousness was so flabbergasting, will never serve the slightest practical or commercial interest. It's not worth a dime. We shrug it off and don't even bother trying to correctly pronounce it.

Ah, but to experience total intellectual humility, just once.

To have been alive on March 13, 1781!

Going to Extremes

On this cozy blue planet, our comfort is so habitual that we only notice it when it's interrupted. A power failure causes us to be temporarily cold, dark, or too warm. But we never face lethal conditions, the kind that are routine in the perilous kingdoms that rule the night's measureless expanses. The brightest, dimmest, largest, smallest, most violent, or least active regions of space-time lie light-years from our front lawns, where they neither impinge upon nor threaten our daily routines. But even at their awesome distances they serve to clarify the nature of things, the structure of the universe that is our home. To understand our planet and our human life-forms, to grasp the larger womb from which we sprang, we should first survey these edges, these extremes, these boundaries of the known universe.

Anyway, superlatives are always fascinating. They're the headline-grabbing cores of news articles and gossip. We can count on gaining someone's attention with reports about the prettiest, wealthiest, smartest, most perverse, or most egocentric individual: Average is much less entertaining.

Like definitions of contemporary morality, some of physical reality's edges are marginal or blurry. Example: Nobody can say exactly what is the diameter of the observable universe. Other limits are not fuzzy at all, but razor-sharp. People are either dead or alive, pregnant or not; the light switch is either on or off. No gray area.

In astrophysics, one of the best defined boundaries is the coldest temperature in the cosmos, at exactly 459.67 degrees below zero Fahrenheit. Since what we call heat is simply atomic or molecular motion, and since all motion stops dead in its tracks at this temperature, there simply cannot be anything chillier. In 1997, distant clouds of gas were found to absorb the meager trace of heat that permeates interstellar space, and such gas clouds became the coldest known places, only a few tenths of a degree above that absolute zero barrier.

While the rest of the universe basks in the 5-degree heat (usually expressed on the Kelvin scale as 2.73 degrees) left over from the Big Bang explosion, the *absolutely* coldest spot, so far as is known, is much closer to home. No, not the Antarctic (where a fairly impressive minus 130 has been recorded)—temperatures within a few *billionths* of a degree of absolute zero exist right here, in ingenious research laboratories. Thus, the coldest place in the universe is found on Earth!

Traveling the other way, upward, there are no limits whatsoever, since temperature has no theoretical ceiling. The sun's surface is a mere 11,000 degrees Fahrenheit, while blue stars, the universe's hottest suns, deliver surface heat of over 40,000 degrees—an energy that has wormed its way outward for years after escaping the stellar core's unbelievable nuclear-bomb temperatures of many millions of degrees.

Even such unfathomable heat is dwarfed by a supernova, typically a *billion*-degree furnace—incomprehensible to our minds but meaningful to the enterprising sizzle of the universe, which depends on such conditions to forge new, otherwise impossible-to-create elements (including the iodine that ended up in our thyroid glands).

If we treated the entire universe as a patient on a hospital bed and kept taking its temperature, we'd find that its present reading of 5 degrees Fahrenheit is only half what it was when it was 38 percent of its present age. The cosmos is getting colder all the time. All the warmth cooked up by its 50 billion visible (and estimated 1 trillion invisible) galaxies—each of which pops off a supernova once per century, on average—fails to make up for the dominant, chilling mechanism of the universe's expansion, like Freon in a refrigerator.

Which leads to our next superlative: speed. This is another easily defined margin, since nothing that weighs anything can quite reach the velocity of electromagnetic radiation (light, infrared, X rays, microwaves, etc.), which buzzes along at 186,282.397 miles per second in a vacuum. Paradoxically, while the fastest objects are perfectly well defined, the slowest moving are not. The reason is that you always have to state something's velocity relative to some other place or reference, and we have no stationary grid system against which motion can be compared. So, a snail on Earth's equator still rotates around our planet's axis at 1,038 miles per hour, whizzes around the sun at 66,000 miles per hour, heads toward the Andromeda Galaxy at 50 miles per second, and makes who knows how many other movements relative to various galaxy clusters.

In our own cosmic neighborhood, the slowest spinning object is Venus, whose equator crawls along at just four miles per hour. A person could *jog* along that planet's waistline and keep the sun from setting. As for motion of an entire body through space, the winner is the slowest-traveling planet: Pluto, at a half-mile per second, needs a quarter of a millennium to orbit the sun.

Speaking of the sun calls up another easy superlative: The most luminous thing in human experience. It delivers 40 billion times more light to us than the Dog Star, Sirius, the night's brightest. The sun is bright enough to cause retinal lesions (bad enough, but not, contrary to popular myth, total blindness) in anyone who stares at it for more than a minute or so at a sitting. Like Lot's

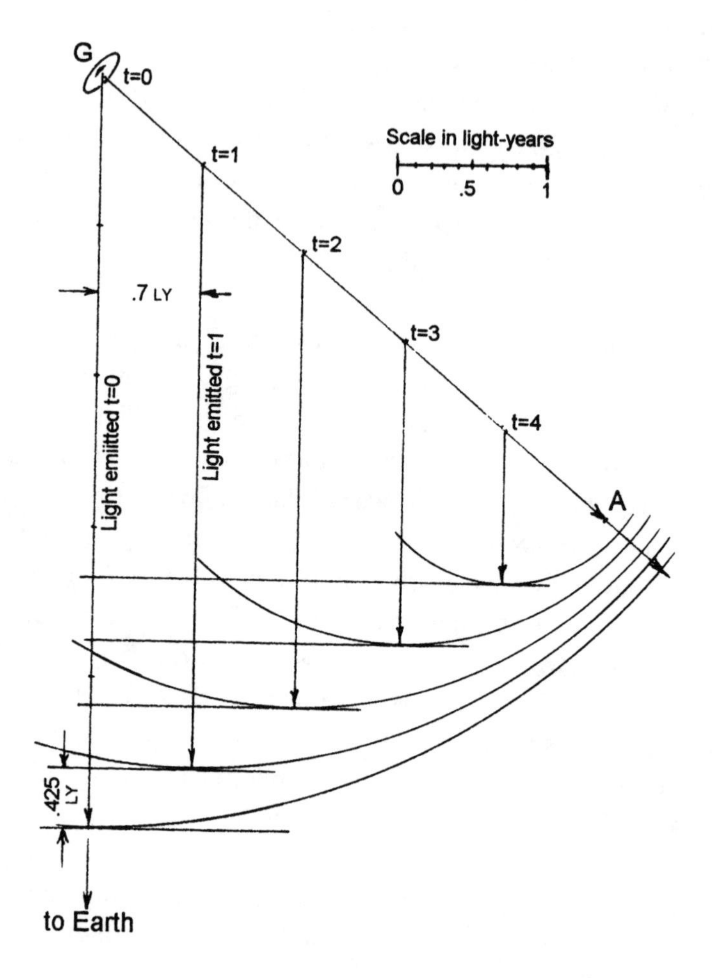

Here's why some objects appear to move faster than light. A bright jet of gas labeled A, traveling at 90 percent of light-speed, leaves galaxy G at time t=0, its light traveling toward Earth at light-speed. One year later (t=1), the jet appears to have traveled seven tenths of a light-year from our perspective, but the light emitted at t=1 is only .425 year behind the light emitted at t=0 because the jet is .55 light-year closer to Earth at t=1. When these light rays reach Earth, they will arrive only .425 year apart, so jet A seems to have moved .7 light-year in only .425 year—1.62 times the speed of light.

wife, we have been forbidden to let our eyes linger on the most fascinating thing in the neighborhood.

Since there are suns as much as a million times brighter than ours, the "most brilliant" award must be presented far beyond our solar system. Rigel, the famous star in Orion's foot, shines with the same light as 60,000 suns. Not impressed? Then how about the amazing and enigmatic star S Doradus, visible from near the equator and all points southward? It has a luminosity of 500,000 suns. Not yet bright enough? In 1997, the Hubble Space Telescope found a star emitting 1 million times the sun's brilliance—the all-time, all-star winner.

But a supernova is the straight flush that beats a full house: It beats any star. In 1987, the supernova in our own companion galaxy, the Large Magellanic Cloud, radiated 500 million times the energy of the sun! In the year 1006, Oriental records tell us, such a temporary object shone some one hundred times more brilliantly than the Evening Star and cast distinct shadows on the countryside. At its estimated distance of 3,000 light-years, it would have had to emit *100 billion times the light of the sun.* And that's the winner in the temporary-visible-light luminosity department.

As for the brightest inherent thing now visible to the naked eye, that has to be an oval, cloudy smudge in the constellation of Andromeda, visible in rural skies every moonless autumn night. This, the Andromeda Galaxy, emits the same light as 100 billion suns. Andromeda also sneaks into a second listing of superlatives: It's the farthest thing generally visible to the unaided eye.

My nominee for most energetic body is any healthy two-year-old. Celestially speaking, though, it's the gamma-ray burster, whose enigmatic energy pops off once a day from distant realms of the cosmos (See the chapter called "Bohdan's Bursters," page 214). In a single second, each burst emits the same power as the sun has generated for its entire lifetime of 4.5 billion years. Such short-lived explosions, which dwarf even supernovas, may be caused by collisions between stars—exceedingly rare events in to-

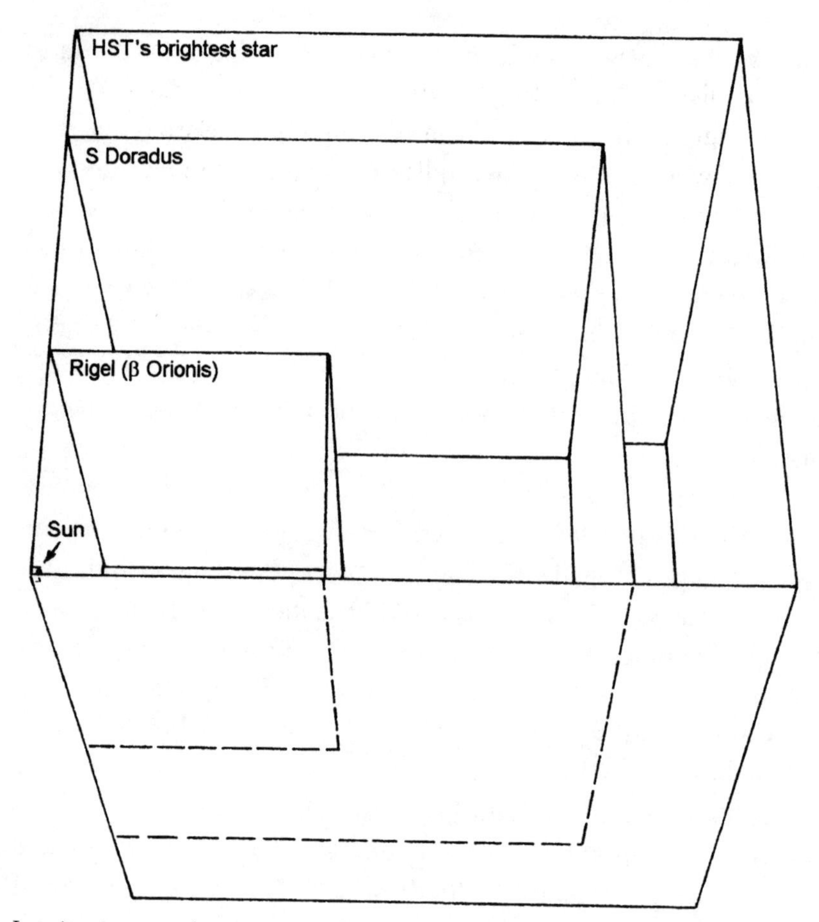

Luminosity at a glance. To picture the brilliance of the brightest star seen by the Hubble Space Telescope (HST), we've constructed these light boxes. The volume of each bin is proportional to the light given off by each star.

day's spacious universe but more plausible way back in the long-ago violent era from which their light has come to us.

In the quick-pulse-of-brilliance department, our final superlative award can come right back to Earth, with the brightest artificial light—the ten-billionths-of-a-second, 1-trillion-watt laser invented in 1995.

Having explored the slowest, fastest, brightest, hottest, and coldest things in the cosmos, now let's leave the spatial dimension in favor of the temporal realm: What are the longest and shortest intervals of time that can carry any meaning? (You've seen in the chapter called "Time and Time Again," page 134, that time can bewilderingly change its duration as well as be relative in other ways, like when my friend Jim tells his long, boring favorite story to a new audience.)

The longest periods known (other than the elapsed time since the present universe was born, some 14 billion to 16 billion years ago) are the rotation periods of galaxies. Since these are recurring, repetitious, cyclical events, they are clocklike; their ticktocks chime off the most sluggish meaningful intervals. If our galaxy is representative, a typical period of rotation is 240 million years.

The shortest possible time period is theoretical but is nevertheless much more precisely delineated. It is the Planck-Wheeler time, named for physicists Max Planck and John Wheeler, who showed that any interval shorter than a ten millionth of a trillionth of a trillionth of a trillionth of a second becomes meaningless, since it would then be impossible to know which event came first and which afterward.

The best candidate for rarest element: uranium. Just one such atom exists for every trillion of the most prevalent substance, hydrogen. Or, looking at it another way, for every trillion hydrogen atoms, you'd stumble across 80 billion helium, 740,000 oxygen, 450,000 carbon, and a single uranium atom.

The smallest objects must be found in the zoo of subatomic particles such as quarks; nearly everyone expects these, in turn, to be composed of yet smaller stuff. Obviously, largest objects are easier to classify because they're easier to see and far less likely to have their status usurped. In this category, red giant stars such as Betelgeuse or Mira take the prize. They're so enormous that on a scale model in which each might be portrayed as a hot-air balloon twenty-five stories in diameter, Earth would be the period at the end of this sentence.

Of course, collections of stars are even larger, so that galaxies become the biggest discrete items known, and the bloated, inflated kings of galaxies are the giant ellipticals such as the awesome M87 that sits at the heart of the Virgo Galaxy Group, some 60 million light-years away. Traveling at the speed of light, it would take a million years to cross from one end of this colossal sphere to the other.

The fastest-spinning natural objects, at least visible ones, are pulsars. The Crab pulsar (a collapsed sun hard as a diamond and as small as Manhattan), in the constellation Taurus, rotates every thirtieth of a second. Other, less famous pulsars have been found to spin hundreds of times per second, with 860 rotations per second the current record holder. It's dizzying simply to imagine living on the surface of such an object. If one could survive the awesome gravity that would crush any visitor instantly (spreading his remains smoothly around the surface like a film of oil, until no part of the corpse stood more than an atom high), the spin rate would still be lethally sickening. Imagine: The stars of the night sky would whiz from horizon to horizon in less than a thousandth of a second, forming unvarying straight streaks across the heavens that would never change in the slightest.

Except for the coldest temperatures (which, as noted, are created in laboratories) none of the other extremes of the universe are found anywhere near our planet. We have been born and nurtured in a protected cranny as far from the edges of violent physical reality as is possible. Someday, however, people may want to explore or even exploit the extremes of the cosmos, and it would be naive and remarkably dangerous to venture out without fully knowing what and where they are.

Such weird, beyond-the-edge conditions may be surprisingly well fathomed by our minds, but let's never imagine that conceiving is anything like encountering them. Experientially, they will always remain as intensely foreign as a vegan steakhouse.

Celestial Compulsions

When I step under the stars in the blessedly black, unspoiled rural region where I've lived since 1972, my eyes stop in the same places. I check out the intriguing eclipsing star Algol in the constellation Perseus, which grows strangely dim every two days, twenty hours, and forty-nine minutes, an Old Faithful of the heavens. There's probably a million people around the world who watch this cosmic "eruption," and few of us bother looking up its schedule ahead of time; the fun lies in being surprised. Nineteen times out of twenty, Algol looks normal. Then one night you go out, and, bingo, it's lost most of its light and the constellation has been distorted into a new, less familiar shape.

Such variable stars have fascinated skygazers for more than three centuries, when they were first noted by mostly patrician European observers with excessive leisure time. Though never set down in writing, Algol's fickleness must have caught the attention of much earlier skygazers, going all the way back to ancient desert-dwellers; its very name in Arabic means "the Ghoul."

But here's the point: Is it routine, a mere habit, that pulls my eyes toward such old friends, as if I were subconsciously checking the water dish of a pet cat? Or is it a compulsion? If we perform the same action every time we return to the same setting—in this case the starry dome—where is free will?

I'm wondering about this on this particular night for a specific and personal reason: When I come back indoors from shutting down the observatory, it is time to see my smiling eight-year-old daughter off to bed.

"Good night, Daddy, I love you!" she says, and then kisses me on the cheek six times. It would be even sweeter if she weren't

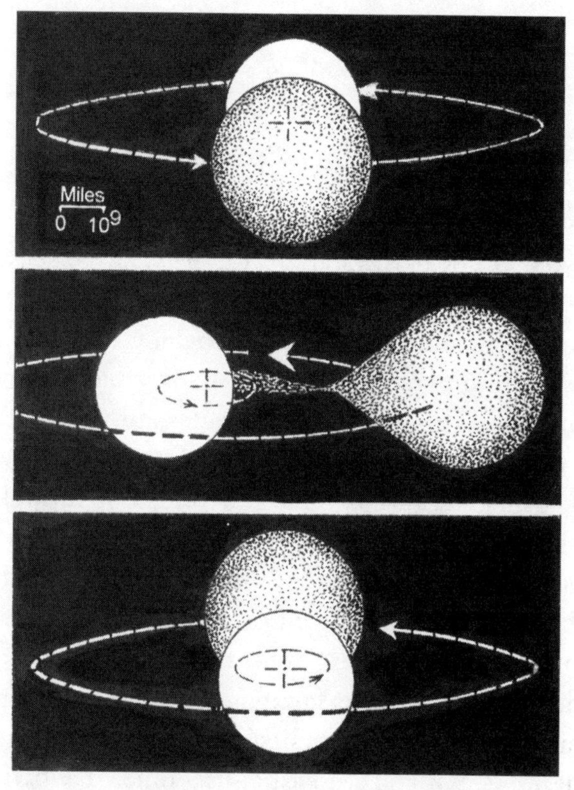

Algol's main stars are too close to be separated by any telescope, but the change of brightness charted at the bottom reveals what must be happening here. As the brighter and dimmer stars take turns partially eclipsing each other, Algol shows obvious changes to the naked eye every two days, twenty hours, forty-nine minutes.

driven to this exact sequence of words and actions. For, as both of us are well aware, she has a compulsion. She cannot kiss me five or seven times, nor can she bring herself to kiss first and then say good night. A psychologist friend shrugs it off and says such obsessive behavior is harmless in the absence of more serious or negative symptoms. Anyway, he has his own compulsions, he confesses, and I look around me, wondering if the whole world isn't a little nuts, after all.

So I make compulsions the topic of the week among my circle of friends and acquaintances. I periodically take surveys like this, asking everyone I know some question about his or her life, something no book or newsmagazine can tell me about contemporary society.

I asked both men and women how many people they'd romantically said "I love you" to in their lives. The survey results were unexpectedly uniform: Most, upon reflection, thought they'd said those words to no less than five, no more than ten people.

Later on, I asked people (including strangers such as waitresses and sales clerks, as well as acquaintances) whether they arrange their money in their wallets (singles with singles, next to the fives, then tens, and so on). And, if they do keep denominations separate and organized, whether they also make sure all bills face the same way. To my surprise, most people, perhaps 80 percent, *do* keep money arranged, and about one in five also keeps bills facing the same way. Since I do not do this, I'd always assumed that only anal-compulsive people would bother organizing their currency. Now it turns out that the money-arrangers are the normal ones; I'm in the minority. It's we haphazard, anything's-okay types who are the oddballs, the comparative slobs.

Anyway, my habit of always looking at the same stars, as well as my daughter's compulsion, inspired me to a new survey—about compulsions. This is the kind of thing you can ask only friends or close acquaintances. Like the "I love you" business, it's too personal to be asked casually.

As I would never have guessed, nearly everyone I asked either still exhibits compulsive behavior or admitted to surrendering to it as a child. One friend, Barbara, said that as a ten-year-old she was convinced that a giant bird would kill her entire family unless she stepped on the bathroom tiles in a particular sequence. For more than three years, she didn't deviate from that pattern. Another friend, a therapist himself, still cannot set the microwave oven unless he punches the numerals sequentially. In other words, to cook something for two minutes, he simply cannot bring himself to press 2:00. No, he has to press 1:23, and when that time's up, he'll add another 34 seconds in order to get close. His college-freshman daughter does the same thing.

More commonly, people say they have to take a certain number of footsteps (for example, four, or four fours—sixteen) to reach the stairs and will stretch out the last few steps to make that work.

It's funny: When you hear about them, other people's compulsions sound nothing short of hilarious. Try it out in your circle of friends some evening if the conversation flags.

While I haven't been aware of any personal compulsions since childhood, what about habitual behavior, like always looking for Algol? At astronomy-club meetings, I've noticed recurring observing behavior in many others as well. Which makes me wonder how much of our lives unfolds by rote, as if we were programmed machines.

One science-fiction movie of the eighties featured people who had "implanted" memories. Everything they recalled as a childhood experience was in reality a recent creation, instilled in their minds synthetically. It was a neat, chilling idea for a subplot, but since our own memories are always highly selective, we really don't do too much better in terms of accuracy. Throw in our routine reactions to daily stimuli, our emotional and intellectual biases, and the fact that habitual behavior is easier and more comfortable than innovative or novel activity, and we're practically androids ourselves. Try to get someone who isn't French to taste snails or frogs' legs, and you'll see resistance to change.

This applies even more rigidly to scientific behavior: What's

"Step on a crack, you'll break your back" causes many wary children to assume an odd, compulsive gait as they walk home from school.

considered appropriate, fashionable, or worthwhile dictates what avenues researchers will pursue. Can anyone doubt NASA's or Stephen Hawking's unintentional influence on astronomy funding?

Does the press, reflecting public interest, latch on to stories about the moon because it is of paramount scientific interest? Hardly. Something primal and yes, habitual, makes that nearest body hold a knee-jerk fascination. We will gaze at the moon and eventually go there again and again because we cannot help ourselves. It's so deeply linked in our collective psyches, we're compelled, obsessed, driven to that dusty ball of sandy stones.

Do compulsions go beyond humans—and perhaps even rule the universe? Every cat I've owned seemed to feel driven to kill birds, even if it wasn't hungry. They'd all chatter their teeth at the sight of a nearby buzzing fly. Where was free will? It seems arrogant to assume that we are somehow removed from the rest of the animal kingdom in this regard.

In leisure-time activities, as well, most of us return to favorite restaurants, order familiar entrées, hang out with established friends. I even know people who book the identical vacation annually. And, of course, it's expected that we wake up and look at the same person year after year. You again!

It seems, then, just one more logical step to act compulsively, to perform identical actions repeatedly.

But tonight I'm going to break free. I'm going to show my students a wide variety of new things. I'm not going to check out Algol. I won't do it.

I won't.

The Vibes of Space

When it comes to the distant universe, things are not as they seem. All telescopes show images of objects from the past, upside down and, nowadays, computer enhanced. Like the artificial waves in theme-park swimming pools or the "virtual reality" of some computer programs, astronomers, too, experience odd paradoxes and distortions of our own creation.

On the one hand, researchers churn out a growing torrent of data that brings us ever closer to the stars. Electronic amplification procedures probe distant galaxies with a clarity that is more up-close-and-personal than ever before. Yet these same devices erect strange, previously unknown barriers between our senses and the glories of the universe. Actually, it's not *senses* (plural) but merely *sense* (singular) that inspects the cosmos—in space, sight alone gets utilized. We cannot touch, taste, hear, or, thankfully, smell any other celestial body. (Thankfully, because there are worlds of sulfur like Jupiter's giant satellite Io that would send us a rotten-egg blast, and also planets with thick, putrid atmospheres of methane—swamp gas—and ammonia.)

Nor do researchers "listen," except in Hollywood movies. Sci-fi routinely shows radio astronomers wearing headsets as if the stars emitted a staticky version of *Prairie Home Companion.* But despite their name, radio telescopes do not detect sounds of any kind. Their parabolic surfaces focus electromagnetic radiation, a form of light that simply has a longer wavelength than our eyes (or optical telescopes) can perceive.

Everything known is first seen. Then the information is analyzed. So our first issue here is: What happens when there is no initial seeing? When mere ideas spawn other ideas, which in turn lead to conclusions, which generate yet further theories—a pyramid whose foundation is the brain alone, without any initial data or the slightest chance of confirmational observation, nor any experiment that can choose between theory A and theory B?

Most of us would be wary of such a scheme. Yet it is precisely what many theoretical physicists are up to as they seek such esoteric holy grails as a Unified Field Theory. (More recently, it's been called the Theory of Everything. This has a nice ring to it. And it's a reasonable progression, since we've found time and again that seemingly disparate forces are really two faces of some larger whole.)

Actually, the move away from direct observations has been steady and increasingly accepted in theoretical circles. Perhaps it started with the abandonment of the telescope as an instrument of direct perception, an evolution that has gone unnoticed by the public. A 1997 survey came up with the unsurprising fact that the majority of Americans cannot correctly locate many of the countries in Central or South America; the world is bathed in deep mystery once Joe Citizen goes farther than the fridge. So it's hardly astonishing that most people are unaware that astronomers no longer look "through" telescopes. That practice, already waning in the late nineteenth century, vanished like a dream in the middle years of the twentieth.

The first process that made humans obsolete at the eyepiece was photography, which accumulated light and produced images utterly invisible to even the sharpest eye at the greatest telescopes. Only one lingering reason to look through a telescope remained: to discern fine planetary detail, because photography's single Achilles heel was its need for time exposures that simultaneously suffered the blurring effects of Earth's ever-dancing atmosphere. A patient observer, however, could sit at the eyepiece for hours, alert for the moments when the air would momentarily steady like a camera suddenly reaching focus. The sharp but transient detail would then be sketched on the spot.

Such drawings at the telescope provided the best planetary information for centuries. As late as 1980, some sharp-eyed amateurs sketched radial spokes on Saturn's rings, though no photograph had ever shown such odd features. Experts were unanimous in dismissing these as hallucinations akin to the spurious canals of Mars; no ring-spanning feature could possibly hold together given the rings' differential rotation. They were patently impossible. But then Voyager visited the ringed behemoth in 1980 and sent back televised images—eureka! They showed those bewildering spoked rays of electrostatic origin. Now the obstinate observers were suddenly elevated from pinheads to prophets.

Ironically, those same spacecraft also brought about the final doom for direct eyepiece observations. Who could compete with such close-up snooping? The additional technological advances of adaptive optics (see page 122) that could electronically remove much of the atmospheric smudginess, plus satellites and spacecraft like Hubble that captured images above our obscuring atmosphere, made looking "through" a telescope an archaic and pointless exercise so far as research was concerned.

Photography's replacement—electronic capture and amplification of the image—also allowed later analysis as well as manipulation and enhancement by computer techniques. Equally significant, much more could be learned by breaking the incoming light into its component wavelengths through spectrographic analysis.

Astronomers analyzed, manipulated, and studied the data at

their home research center or university. There was no need to be up at night or even to be physically present at the telescope. Indeed, telescopes themselves were increasingly being commanded and controlled from afar—sometimes by researchers continents apart.

All this made for ever-easier research, aided and abetted by the Internet, which allows instant conferences and communications among those engaged in any particular area of study. Where a mere twenty years ago many were still trying to squeeze greater data

The bizarre "spokes" on Saturn's rings are evidently shadows cast by thin clouds of dust elevated above the ring plane by Saturn's magnetic field, which is ten thousand times stronger than Earth's.

from improved photographic techniques, nowadays even the photographic plate is considered archaic. A photograph cannot enhance, twist, change the contrast, bring out subtle highlights. Nor can it transmit to remote monitors an image of the broken fragments of a violent galaxy so faint that its very existence would lie beyond the limit of the human eye at even the largest telescope.

It figures: As knowledge gushed forth like water from a broken dam, professional astronomers became ever more removed from the actual night sky.

I watched one day as the chairperson of a major university astronomy department looked at the image of a galaxy that had just arrived from the Hubble Space Telescope. She shrugged when asked what constellation contained that faraway city of suns. "Let's see." She squinted and, giving up, instead read off its celestial coordinates. Point: It didn't matter to her. The lessons learned from the stars being born in this achingly distant Milky Way lookalike had nothing to do with knowing where it lurked in the night sky. It was as irrelevant as if a medical researcher had been asked what color was the guinea pig being injected with an experimental drug and where had it spent its childhood.

Meanwhile, however, on the other side of the night, an army of worldwide amateurs continues to inspect the heavens. Many of these people know the sky the way the ancient Greeks or Arabs did—as well as the rest of us know the pattern of ceiling flaws above our beds. For them, the yearly return of Leo affords a peculiar thrill, and an exploding meteor ripping the heavens in two is reason enough for sacrificing a few hours of sleep.

These people are aware, even though few probably put it into words, that there is a big difference between image and actuality. Just as a video recording of a birth doesn't begin to capture the flavor of the real event (awesome! I was there!), *images* of celestial wonders are a far cry, experientially, from the real McCoy.

A meteor depicted in a movie is an insipid streak that hardly raises eyebrows in the theater. An actual "shooting star" brings involuntary gasps to all who see it. Why is this? Who can explain why the sudden sparking path of a fragment of stone no larger than a raisin should excite us so?

A rainbow on television is a caricature: flat, hollow, absent of any grandeur. An actual rainbow is more than vibrant; it's exalting, almost otherworldly. To those who have been to the Grand Canyon, then see it later on television, this dichotomy is obvious. The whole essence evaporates in the conversion to electrons and glowing phosphors. Much more than the all-important sense of depth and the panoramic, wraparound experience has disappeared—something else is gone as well. In most cases, people are aware of the huge split between image and actuality, but some of today's theme-park rides, IMAX, and virtual-reality games are beginning to blur the difference.

So it is with space. Everyone who has seen a total solar eclipse finds the experience overwhelming. More than just animals go berserk when the moon covers the sun and its normally invisible atmosphere leaps across the sky like some alien kaleidoscope. Something visceral happens when the sun, moon, and your spot of Earth form a perfectly straight line in space. Yet photographs of total eclipses make the event seem barely intriguing, merely "interesting." Photo versus being there is visually the difference between "pleasant" and "life's mightiest experience."

On a lesser but still impressive level, comets also carry a palpable presence, as do the northern lights. The Milky Way as seen from desert-quality environs is another vibrational powerhouse: The ancient Peruvians considered it the heart of all existence.

Moving from such grand naked-eye sights to those seen through binoculars or telescope, we find that their impact doesn't seem reduced by the introduction of lenses or mirrors. (We might try such an experiment by looking at a person, then at that person's reflection in a full-length mirror. Does the person's essence or gestalt seem as palpably present in the mirror?) It's almost as if the "vibes" of the celestial object remain intact when the object is reflected or focused by optical surfaces.

Example: Saturn. Anyone who has ever seen that ringed world through a good instrument will easily recall his or her reaction. As director of two observatories, I've watched thousands view that surrealistic planet over the years; the same two expressions always recur. People gasp and say either "Oh my God!" or "That's not real!"

This is curious. What are people unconsciously expressing when they say, "That's not real!"? I think it's simply that the real Saturn carries an authenticity, a presence, an overwhelmingly beautiful essence that is very different from a photograph. Indeed, most have seen photos of Saturn far superior to the views through our observatory instruments. Yet only the latter experience produces the verbal pyrotechnics. It's the image versus reality thing, once more.

There is an unfortunate trend in observatories to use electronic imaging and to display the object on monitors. This indeed captures the image: There is Saturn, sitting flatly on the television screen. But its essence has been completely filtered out; its life is gone. It's a postcard of the Taj Mahal, a 900 sex call, a symphony heard through a Walkman, a hot tub instead of a tropical lagoon, a poster of a beautiful woman, a supermarket tomato, a *description* of a shiatsu massage.

And that is what modern astronomy delivers. It is food for the intellect but not for the soul. It provides more answers than ever before, even as it opens up twice as many questions.

If its goal is to expand our minds and increase our sophistication and depth of knowledge of the cosmos, it is wildly successful. If it is to competently catalog five thousand galaxies where a century ago we knew of only one, we're on the right track. Our knowledge is vastly more cosmopolitan, and, as if deliberately drawing on that word, we are farther along in our understanding of the cosmos.

But if we are trying to give more people the feeling of the glory of the cosmos, to provide a taste of the inexpressible grandeur of the denizens of the deepest deep, then the sidewalk astronomer and local astronomy club with telescopes set up for public use provide the better service.

For the telescope is the end-stage device that brings home the "vibes" of space. Any further efforts at enhancement or improvement accomplish just the opposite. Like the thick clear plastic my grandparents kept over their furniture (reducing potentially comfortable chairs to sweaty, crinkly intimacies with packaging material), these electronic membranes bleed the heavens of their power. Sensation has dissolved; image alone remains. Like the wax replicas in Madame Tussaud's museum, caricature's of life's animation, the monitors increasingly evident in public-use observatories are well-intentioned but ill-conceived efforts to allow the public to "view" the heavens. The televised star cluster is frozen on the screen, while the visitor, led along as if on a Disneyland ride, files past—as removed from the quirky pulse of life as a deer's head on a barroom wall.

So what is this "vibe" business? How is it that celestial objects can have "vibrations" or an "essence" or "aura"? Coming from someone who finds New Age nonsense intolerable, the question may seem thoroughly out of place. Yet it's undeniable. When the Indian government gave me time on their 40-inch telescope at the Nainital observatory in the Himalayan foothills, we set it up to view Saturn visually. The staff astronomers and I had all seen the ringed world hundreds of times before, yet, as it hung suspended in such steady atmospheric conditions as to look like the New Year's Eve ball before it drops in Times Square, we each leaped up and clapped for joy. A color transparency does not produce such a reaction. Only the real thing inspires—and it's been that way since humans first gazed upward.

I'm not about to set myself up as "investigator of vibes," though such an occupation would probably be enormously successful in my hometown of Woodstock. Suffice to say that people do seem to be able to "sense" things. We walk into a room where a quarrel has occurred moments before, and we just don't like it.

Something feels wrong. "Bad vibes," we say afterward. Or we stand next to someone on the supermarket checkout and he or she just "feels" good. Some sort of peaceful but energizing aura simply seems to surround that person. Photos would not let us "feel" the vibes. It's something else. As is said so often, "You had to have been there."

There's no use cataloging and characterizing the many vibes and feelings we get as we stroll through life. We're accustomed to them and react on some level without thought. The same is true of space. Like a vivid dream that influences the dreamer's mood even after awakening, our model of the cosmos colors our feelings about everyday reality. If we knew that the universe will escape into another dimension as surely as a pet hamster, or if someone absolutely proved or disproved the existence of God, our lives would be flavored by that larger picture.

There is something about the sight of endless points of light in the night sky that stirs the spirit. The mind is numbed by the wide expanse of sky in a dark rural setting. We may find ourselves telling a friend the next day that we felt small, but that's not really accurate.

A moonless night under the starry sky, the real thing, far from city lights, does not make me feel small. Or large. Or anything at all. It produces an exhilarating emptiness. The maniacal riot of senseless patterns punctured by the unexpected meteor creates such muteness that I can feel the night wind rush through me as if I were spirit material. Not small. Nothing.

When we truly feel empty, suddenly we are as everything. At such a time the universe has fulfilled its apparent mission of wondering at itself. And this, very possibly, is the sky's greatest gift.

Going Places

Earth, to some, is getting a bit stale, what with MTV and freeways. And we've all suspected that there's a whiz of a good universe next door, somewhere. Imaginative sci-fi movies such as *Forbidden Planet* often portray worlds of intriguing experiences; on that particular planet, a chairlike machine invented by a race called the Krell allows you to boost your IQ by a factor of ten—a real temptation on sluggish mornings. But why travel farther than we have to? Why move into a distant, run-down house when we can renovate one that's next door?

Could we transform Mars into an Earth-like paradise? Could that one-time vacationland be resurrected? Ample evidence from the visiting Viking and Pathfinder space probes shows that Mars was once a pleasant place with moderate temperatures and abundant flowing water beneath a thick if oxygen-poor atmosphere. It's only half the diameter of Earth but, with no oceans, possesses exactly the same land area as our own world. Mars, a shadow of its former self, has nonetheless enjoyed good press for more than a century as the most inviting planet in a rogues' gallery of deplorable choices. It cries out for more than mere exploration. It invites *renovation*—a transformation to an Earth-like state.

The idea has crossed the line from science fiction; nowadays such dreams of *terraforming* draw speculation from serious-minded people. What a wonderful notion: The cosmos teems with valueless globes, each as barren as a political promise.

The concept seems simple enough. You ignore hopeless cases like Venus or Pluto and instead zero in on a world like Mars that already has some of the essentials (ice below the surface, oxygen bound into the soil). You "seed" that planet: You add something that serves as a catalyst, which sets off a chain reaction. The process might take centuries, but eventually you end up with breathable air and voilà: Eden Two. The formula has become one of the utopian staples of science fiction.

But the reason it's so appealing, I think, goes beyond the gung-ho spirit or Yankee can-do ingenuity. It even lies beyond the endorsement of outward exploration that's been ingrained in most of us since elementary school. The tacit cosmic Manifest Destiny in this idea may be a mixture of arrogance and a Puritan work ethic that suggests we'd be guilty of lazy neglect, even slothful, if we don't harvest all the real estate we can. And if we can turn a barren, dilapidated, outer-space vacant lot into a prime plot, an ambrosial landscape with a dreamy view of mottled, spinning moons, well, why not?

Here's why not. Compared with our world, other planets have disheartening differences integrated into their very core. Mars is an excellent example, since its similarities to Earth make it a more compatible target. But there are many perils in the equation. Let's start with Martian gravity: just 38 percent of ours. Which means any thick, precious, Earth-like atmosphere we created would leak into space like water through corroded plumbing. Mars also receives less than half the sunlight we do. And it has no oceanic area to moderate seasonal extremes. Such are the basics we can never control. But it gets worse.

With the rocketry equivalent of a caravan of U-Haul trailers,

A key part of terraforming, at the mouth of Ares Vallis on Mars: the erection of real estate signs.

we could drag machines 36 million miles to extract oxygen and water from Martian raw materials and thus supply a bubble-enclosed (and necessarily limited) colony. But imagining that it would transform the entire atmosphere ignores the volume and scale implicit in global meddling. It would be like expecting a bathroom heater to keep North America toasty in winter.

Nature allows few wildly unstable situations, where the introduction of a tiny new factor changes everything radically. True, you can insert a new parasite into an environment and suddenly all the maples are gone. Or let a few rabbits loose in Australia and find the country overrun a decade later. But return two centuries beyond that, and nature has found a way to cope with the situation. The natural state rebalances: The whole planet does not experience radical transformation. Even something as drastic as an ice age produced by orbital changes or the uplift of a major mountain chain may seem extreme but is actually minor when considered in the context of all global parameters. Of course, the birth of the Himalayas permanently altered our own global climate and probably changed the North Pole from a balmy swamp where dinosaurs munched greenery to the abandoned frozen monochrome of today.

But even so, the real basics didn't budge. For example, atmospheric nitrogen and oxygen did not switch over to methane and hydrogen.

With terraforming, we're asking for something far more radical. We're imagining that if we introduce a few thousand tons of a new substance, it will catalytically cause the entire planet to undergo a self-sustaining chain reaction whose end result is Earth-like.

So the negatives are far more prosaic than deliciously scary nightmares of underground Martians emerging to wipe out our human colonists like Apaches circling the wagon train, or imaginative but more plausible fears of some fundamental but previously unknown human physical need (like a magnetic field) going unmet in the Martian environment.

Terraforming also happens to be a dangerous idea on an ethical or philosophical rather than technological note: It suggests that we may create alternative habitats to Earth, making this home planet somewhat expendable. In this almost futuristic era of virtual reality, the inescapable reality is this: No other place anywhere in our solar system will ever offer the songs of robins at dawn or the opportunity for children to see whimsical figures in puffy afternoon cumulus.

Sitting in the stark, alien beauty of the Persian desert near the town of Kerman, I once felt so horribly isolated that I grasped a truth known to every lonely seagoer from Vikings to modern submariners: There are no jewels as precious, no security as soothing as home. And we have but one. We can cherish it or we can ruin it: our choice. Shall it become orderly "tree farms," where unvarying rows of a single arbor species stretch safely into the distance, or instead retain forests, where things remain wild and unexpected? There are no other places to live, and there never will be, save in synthetic environments. Terraforming offers just that, promises of tedium, artifice that would make Disney World seem a wilderness expedition by comparison, even if commercial interests found ways to exploit features like that pink sky and weaker gravity to create amusement parks with innovative thrills.

Science fiction is fond of offering intriguing concepts as innovative and varied as the fevered minds from which they sprang. In recent decades, we've been presented with ideas like these:

- bending space so that astronauts could "jump" to other sections of the cosmos without passing through any of the intervening territory
- creating wormholes to "tunnel" back through time
- terraforming planets for easy human colonization

Simple to grasp, they suffer a common practical shortcoming: Nobody has yet come up with the slightest clue as to what could possibly produce such effects. Or how we could even begin the process. Like the perpetual-motion machine, some ideas have no analog in real life. Terraforming may (or may not) be one of them.

Meanwhile, we teeter on the edge, approaching a time when we may lose the ability to be self-sustaining here at home, thanks to excessive human population and the understandable desires of three billion Third World inhabitants to reach Western living standards. When everyone on the planet has a car, there may be few beautiful places worth driving to. If everyone owns a television but cultural diversity has disappeared into a worldwide ersatz Hollywood, what will be worth watching?

Not that long ago, real cultural differences existed. Tribes of naked people were as comfortable on this planet as were Eskimos in igloos. In the late sixties, it was fascinating to find John Kennedy's photograph in a remote Indian village, or to see wild panthers when riding on the back of an elephant, and routinely to come across people who were genuinely afraid of having their photographs taken. Back then, television simply didn't exist in most of Asia, and bizarre but fascinating customs ruled everyday life. Now it is almost invariably disheartening to visit a place you've always thought to be exotic and intriguing—say Bangkok, Bombay, or Istanbul—and find a large, bustling, Western-style city with VCRs, discos, McDonald's, and nothing more than variations on the same culture you left behind in Cleveland. Why leave home?

So, sure, let's terraform, if we can ever figure out how to do it. But before we create another Los Angeles 40 million miles from New York, with a longer commute, let's first reexamine what that hackneyed phrase *quality of life* really means. We might *plan* our extraterrestrial experiences, the way the city fathers of Paris and New York planned their beautiful parks and aqueducts, rather than shortsightedly replicating a fun house of mirrors to be encountered again and again no matter how far we rocket. Once we fully grasp the perils of exporting the status quo to other realms, and creatively allow each world to evolve according to its own uniqueness, then might terraforming offer a worthwhile first annex to an over-crowded planet.

Rice Is Nice: New Age Reasoning

How do you react when you read about a seemingly preposterous new discovery? We all draw a line that lies somewhere between open-mindedness and gullibility. The exact location of that boundary critically determines what we're willing to accept as truth. If we're too closed-minded, we fail to be receptive to radically new ideas—which is why plate tectonics and sudden mass extinctions caused by asteroid impacts were so thoroughly dismissed when they were first proposed. Conversely, believing every new wacky idea without proper analysis retards the search for knowledge just as effectively.

For me, the arena for the most intense battle between naïveté and skepticism was the West Coast a few decades ago. A strange series of events had brought me to San Francisco just in time for the hippie movement in 1966. I doubt that anybody who was there has ever been the same. The two years I lived there crystallized my involvement with science and gave me firsthand instruction in New Age thinking.

It started when I "eloped" with my parents' housekeeper, a

beautiful, blond, twenty-three-year-old Englishwoman three years my senior, whose obsessive desire to see the American Wild West eventually led us to the fabled city of seven hills.

My English girlfriend ultimately returned to the East Coast, but I lingered, mesmerized by the fresh, philosophically liberal mindset of the young longhaired people who were just starting to flock to that city during the magical summer of '66. There were things to see and hear and experiences to share that might as well have occurred at a youthful hangout in ancient Greece, or anytime, anywhere in history that a gust of fresh air stimulated an entire generation.

Typical scene: At an acquaintance's home, intricate designs had been painted onto every square inch of floor and wall. It was a sensory riot, but one's eye immediately went to a single sharply outlined 8 × 11-inch rectangle of bare virgin wood in the middle of the floor. It took a moment to realize what must have happened. The artist, probably LSD-influenced, had started with a sheet of paper on the floor and then kept painting outward until the entire house was covered. Eventually the paper that started it all was removed or lost.

Nothing was sacred, everything was new. The book *Zen Macrobiotics* was all the rage. It had nothing to do with Zen; its dietary counsel boiled down to advising a diet of pure brown rice alone. The reasoning seemed sound, the health and spiritual benefits logical. I tried it for a couple of weeks. Brown rice and nothing else except the salty soy sauce you were allowed to use for flavoring.

A fortnight later, I was horribly ill. Demonstrating my own peculiar inability to see the obvious, I went to a physician and complained that I had no energy and could scarcely move and felt unbelievably terrible. At around his fifth or sixth question regarding my history and lifestyle, he uncovered the fact that I ate only rice.

"Well," he said, rubbing his chin, "rice is nice, but you really need lots of other things. Eat something else and see what happens." Provisionally accepting this insight, upon leaving his office

I promptly downed a cheeseburger and almost immediately felt surges of energy and joie de vivre coursing through every vein. When has a cheeseburger ever had such a salutary effect?

Later I met other "macrobiotics," and they were always skinny, sickly, and humorless. With little reluctance, I realized that the author of this self-help book was a nincompoop. I kept recalling the physician's timeless wisdom ("Rice is nice, but . . ."), and even today I can't sit in front of a plate of basmati or Uncle Ben's without "Rice is nice" echoing in my mind like a loathsome oldie that repeats, unbidden.

In retrospect, it would seem that the most natural diet would be one that humans encounter in nature, where fruits and nuts fall from trees in front of us and we are sensorially attracted to their taste. Shazaam: human food. Going one step further, in cold climates, where fruits and vegetables do not grow throughout the year, it would seem logical that humans might hunt and eat animals to stay alive. But the act of growing and then harvesting grain, and then needing fire and boiling water to prepare it—surely that's another step from being strictly "natural," though such foods seem agreeable enough once they're cooked. In short, I don't know why I bought into the idea that a single prepared foodstuff should be consumed to the exclusion of everything else, but it was only one of a galaxy of irrational concepts that were making the New Age rounds.

Astrology came soon enough. As an avid amateur astronomer, I immediately saw that astrology's signs of the zodiac had no correspondence with the actual constellations of the night sky. The first of many heated arguments with astrologers occurred in a living room on Downey Street, where the astrologist, pointing to a page, insisted that Mars was in Gemini and refused my entreaties to look out the window, where Mars shone placidly amid the stars of Taurus. To this day, the fact that astrological signs do not match up with the real night sky is used by skeptics as an antiastrological argument. (Actually, there are other, much better ones.)

Years later, crystals came into popularity. Supposedly those blocks of quartz had healing or curative powers. One evening, a group of friends admired a huge, magnificent specimen someone had brought, and a few noticed with amazement that they could "feel" its "energy." One particularly vocal Frenchwoman turned to me scornfully and, in the style that seems perfected by the French, sarcastically condemned my ignorant skepticism.

"What's wrong with you?" she sneered rhetorically in that accent that has bewilderingly become associated with sophistication. "Are you completely insensitive? Put your hand over the crystal. You can *feel* its power!"

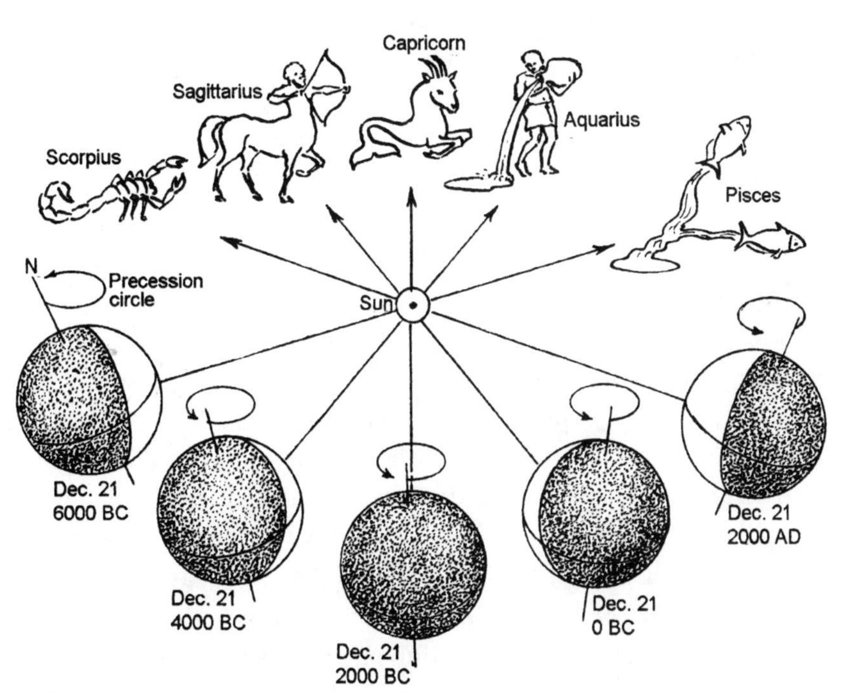

Earth's axis precesses like a top's, arcing through a complete circle every 26,000 years, which causes the sun's position at a given date to migrate slowly through different constellations. (The winter solstice position is depicted here.) Astrology in India and the Orient keeps track of these changes, saying, for example, that the sun is "in" Aquarius in late January. But astrologers in Europe and the United States have stuck to the traditional "signs," placing late January births in Capricorn, as they were in ancient times. Thus, Western astrology does not follow the actual constellations of the night sky.

I argued in vain that when you believe something, your mind can make it happen. "Concentrate on a tingle in your right foot," I pleaded, "and it will materialize almost immediately." No dice. Nobody believed the crystal was anything less than magic.

"Okay," I said, hitting on a desperate idea, "let's try an experiment."

That's how it came to pass that each person agreed to be blindfolded in turn and have his or her hands held a few inches above, sequentially, a telephone, the crystal, a book, and an apple. Could they feel the crystal's tingling effect now, in this multiple-choice test? Five people tried. Only one correctly identified the crystal—just what you'd expect from the laws of chance. Happily, the Frenchwoman was not the lucky winner.

Now, from all this, did I get praise or acknowledgent that science is superior to superstition? Could I even have gotten a date with the Frenchwoman? Not a chance.

One of the tacit themes of the hippie movement was an awareness that *there's more in heaven and Earth than is dreamt of in our philosophy*, and that was good. What we know is as a single snowflake in a blizzard of what is still to be learned. We *must* be open to new, sometimes bizarre ideas if we're to progress and not stagnate. But being open to scams, twisted logic, biased statistics, or assumptions that truth always lies in the opposite direction of what is false does the knowledge quest no justice either. Believing everything we hear is as counterproductive as believing nothing, and I soon realized that the greatest New Age error lay in assuming that every system or set of beliefs that stood outside the mainstream establishment must be valid. If our parents disbelieved it, it must be genuine, and if they believed in something, then it must be false.

We'd been told too many lies about too many things (including Vietnam), and now we were ready to embrace numerology, palmistry, astrology, and all the rest, without proper examination. If our schools and our scientists said these were false, well, that was enough to endow them with credibility. It was so simplistic; it jibed with our uncomplicated credos of peace and love.

The one lingering New Age belief that I still cannot completely shake is ESP. I know and honor the rationale: We preferentially recall those times that someone says something just as we're thinking it and ignore all the many times when this does not happen. Eventually a belief in ESP develops, and it's reinforced whenever such coincidences recur, as indeed they must, given the law of averages. Yes, this is the explanation, and it's a good one.

Trouble is, the synchroneity of thoughts, images, and ideas appears, even to my skeptical brain, to happen far more than can possibly be explained away by chance. Case in point: Two close friends of mine are identical twins. Both are involved in the sciences—one holds a doctorate in biology. They insist that throughout their lives, one would be singing a song entirely in his mind when the other would suddenly burst out with the same tune—in exactly the same part of the song and in the same key!

I, too, have had similar experiences. The odds of such a wildly unlikely event happening are so extremely small, its recurrence defies probability and begs for another explanation beyond coincidence. So who knows?

One thing is certain: It's extremely easy (and fun) to fool people, and potentially profitable as well. One Indian guru who attracted a huge following of largely professional people not only charged lots of money, had his disciples wear his picture around their necks, slept with many of his female followers at will, and had his ashram guarded with machine guns, but purchased a fleet of ninety-two Rolls-Royces—all paid for by devotee contributions. Somehow, few of them found anything amiss in this arrangement.

Yet India, that bastion of superstition, is also an increasingly influential computer programming center and has not neglected astrophysics, either. When invited by the Indian government to use their largest telescope, in the Himalayan foothills, I found the experience amazing and enlightening, first because most of the astronomers walked on footpaths from their homes in a nearby town, where, until very recently, tigers had been seen, and also because the surrounding presence of the Himalayas made the whole thing particularly otherworldly.

From the still, rarefied air of that observatory, stars and galaxies leaped at the observer with a rock-steady presence. It was like stargazing from the moon. On one trip to India, in 1986, I brought a small group with me to observe Halley's Comet (you had to get close to the equator to see it at its best). Unfortunately, in keeping with the unpredictable nature of comets, Halley's tail vanished like a lizard's just as we arrived, leaving the famous visitor looking no better than a bright smudge. (We wished there was some cosmic address we could write to for a refund.)

Indians who are modern enough to study astronomy are as antinonsense and unsuperstitious as physicists in any Western country, their skepticism contrasting starkly with the ingenuousness of their next-door neighbors in the village. But it was those latter, innocent, everyday people we'd usually encounter—people whose naïveté, though delightful, made them believe everything unquestioningly.

My hobby is close-up magic, and one man in our group was a professional stage magician, so between us we'd often entertain our Indian acquaintances with baffling sleights of hand. To my surprise, we were then suddenly worshiped—revered for having spiritual "powers," our protests to the contrary dismissed. They were sure we were just being modest in insisting that we were not genuine gurus or sorcerers.

The potential to exploit the gullible is almost infinite, as magicians have always observed. Sadly, the sciences have not been immune to the temptation, and it's amazing that the hoaxes of the nineteenth century (when, for example, newspapers reported in earnest detail that plants and animals had been observed on the moon) have not been repeated more often in our own era.

Still, a 1996 tabloid headline announced: VANISHED BERMUDA TRIANGLE AIRPLANES SEEN FLYING NEAR MARS. Such "news" can

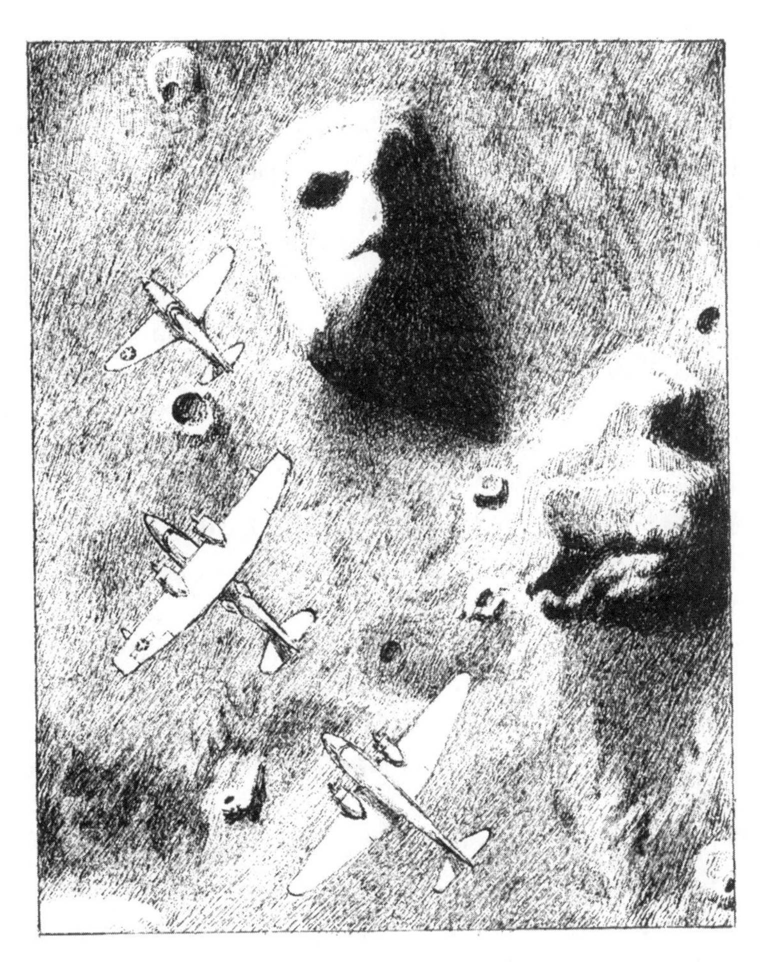

Three Bermuda Triangle planes fly over a Martian landmark made famous by its coincidental similarity to a human face. The resemblance vanished altogether in high-resolution photographs of this same Cydonia region taken by the Mars Surveyor spacecraft in 1998.

find a receptive audience because, alas, a percentage of our citizenry is clueless about basic science. *The New York Times* published a survey showing that many Americans don't know that Earth orbits the sun and a majority cannot define the term *electron*.

As soon as I got my first satellite dish, I briefly immersed myself in the world of cablesque TV. While PBS and Discovery had some interesting programs, other stations' make-believe documentaries were breathtaking in their polished deceit. Why do most school systems not provide instruction in Basic Gullibility Avoidance?

When I visit schools, kids often speak of UFO abductions as factual because they "saw it on TV." Neither home nor school has let them in on this obvious little factoid: Television shows are not there to tell the truth or to educate. They are there to make money, to sell advertising spots. There is an open-ended market for spooky, vaguely threatening themes and a lucrative demand for suggestive (if always blurry) UFO video clips and "eyewitness" accounts.

After a lecture, a woman came up to ask my opinion of a "miracle" she'd seen. She'd just returned from Medjugorje in the former Yugoslavia, where crowds from around the world have gathered for more than a decade because a group of local teenagers claimed to have seen the Virgin Mary. Well, I'd gone there in 1986 and didn't believe a word of it (and neither did my wife, who had been raised as a Catholic), but, okay, we're talking faith rather than science here and there's no right or wrong. More to our present point, people now there are told that they themselves can participate in the miracle: Anyone who simply stares at the sun for several minutes will see "dancing lights."

This woman had been wowed by the "lights" and wondered if I had any possible explanation for such a miraculous apparition. But she quickly lost interest when I mentioned afterimages that appear when fixating on any brilliant object and once I voiced concern over the dangers of such sun-staring.

The funniest incident regarding supernatural lights happened in 1971 in India, where I'd been touring by motorcycle. One day, I visited a friend at the ashram of a chubby boy who, along with his entire family, claimed divine powers. The big deal and main entrée at this place was an initiation ritual whereupon you could supposedly experience the "clear light." It was all the new arrivals would talk about: when they would finally be "given the light."

Curiosity got the best of me. I'd had an astounding but ineffable experience involving a dazzling light some years earlier (an unexpected revelation that had ecstatically burst upon me while I studied for an exam at college, of all occasions), and I was definitely receptive to a repeat performance. So now I attended days of tedious indoctrination meetings at this strange ashram until finally judged "ready" for the initiation ceremony. The lights were turned off, and the swami walked around the room touching people one by one. Gasps of delight came from the darkness around me as people were "initiated"—and my turn was coming!

Finally the orange-clad priest stood before me where I sat, and then he pushed my eyeballs inward with great pressure. Naturally, along with feeling the pain, I saw "lights." That was it! That was the whole deal! I left, laughing out loud.

I still patiently watch the sky, and like astronomers everywhere, I never see unexplained lights. But I'm still receptive.

Anything will do.

Improving on Perfection

"When a pickpocket meets a saint, all he sees are his pockets." So goes an old Indian proverb, and it's certainly true. Everything on Earth and in the heavens reflects where we ourselves are coming from. The ingenuous, honest people I know think that everybody's basically honest. The cheaters think that everyone's out to cheat them. If you're cynical, the world seems to reinforce your distrust. When you're loving, you see love all around you.

Attitude: It applies to how you view the stars as well. Sadly, research too often becomes bureaucratic, political, or tedious, perhaps mirroring our habits, our culture, and especially our entertainments, which reinforce the notion that life is superficial.

I think one solution is to return to the perception of everyday life as an epic.

This feeling still exists among the world's simpler peoples. In rural Sri Lanka, I noticed that almost nobody is in a hurry, because even familiar experiences are clearly cherished. There, people meet in a quiet and focused way; even the conversation with a fruit vendor becomes a joy. The encounter is savored because there's a sense that the here and now is too precious to be rushed.

With such a mind-set, how can an errand be a chore? You never know whom you will meet, how things will look in the sunlight, whether a sudden wisp of warm south wind will touch your face, or if your dog will suddenly stop pulling and finally heel. It's all a surprise.

That's why I love going to the Southern Hemisphere. I can look to a large swath of sky where I don't know the names of most of the stars and in some cases am unaware whether a certain star pattern belongs within the boundaries of one constellation or the next. I can gaze at the firmament anew, the way I did when I was six. In the feral sweep of the far southern Milky Way, I return to the source of this love affair with the night sky and see the heavens purely once again.

The southern sky offers other treats as well. Only near or south of the equator can one see the closest star, Alpha Centauri. The Southern Cross, that smallest but most brilliant constellation, is there as well, along with the finest star cluster in all the heavens, the dazzling Omega Centauri.

I think part of the everyday challenge in acquiring an exhilarating outlook is to trick the mind by pulling the rug from under its tendency to trivialize; to undermine its habit of misperceiving the extraordinary as the usual.

Situations are rarely novel, neither are the roads we travel in our daily agendas. It's easy to see life as repetitious when everyday reality involves identical stage props and the same repertory cast of characters.

The key step to a merry, adventurous life, as I see it, is to focus on the here and now, on the changing kaleidoscope of current scenes and events. In an elevator recently, the Muzak from the hidden speaker was blaring a familiar oldie, "Limbo Rock." I said (with a straight face) to the stranger next to me, "It's playing to remind us that we are now in limbo." He smiled politely, as if humoring a potentially dangerous weirdo. I enjoyed that as well, for the man's wary stodginess seemed straight out of central casting. But most people will play with you, and these interchanges spark life with fun and freshness.

Many people now teach their kids not to talk to strangers; in my book that's misguided and paranoid. And plain wrong—fear and suspicion of others will deprive them of a lifetime of adventure and delight. We can safeguard our children by teaching them never, ever to be persuaded to *go anywhere with* strangers, no matter what they say or offer. But not to *talk* to them? What kind of future society are we creating?

It's imperative to realize that every situation is really new, and

The southern sky is an invitation to wonder. Omega Centauri, the finest star cluster in the heavens, appears to the naked eye as a smudge just to the right of the Milky Way. The Southern Cross is embedded within the Milky Way just above the statue's forehead, while the Magellanic Clouds are the two smoky wisps at the left.

here's where outer space comes in. Many advanced amateurs have told me that they look at the same astronomical objects again and again and badly need to "get into" something different. And it's true: Beginners first imagine that the number of interesting telescopic targets is as limitless as the universe itself—but then get stuck observing the same handful. In reality, most of the thousands of observable nebulae and galaxies appear as uninspiring fuzzy blobs—hardly superstuff to wow someone with.

Realistically, the number of visually exciting astronomical targets is surprisingly small, explaining why so many telescopes end up stored in the attic. Serious observers might find a couple of hundred such amorphous targets, but for most nonzealots the truly entertaining objects amount to no more than a dozen.

Among them, Saturn is always spectacular when atmospheric conditions provide a steady image. Throw in the moon with its myriad craters and mountain chains, and Jupiter's belts and satellites, and you have the reliable crowd pleasers. Add a dozen colorfully contrasting double stars, star clusters, and nebulae for deep-space variety. Daytime, no sky show should neglect Venus as a crescent or fail to show sunspots through a properly equipped instrument. And that's about it: the universe, distilled into the same repertory show of a few dozen performers, over and over again.

This is a no-brainer problem. The solution is to focus on novel *aspects,* and suddenly, viewing prospects multiply dramatically. Seeing these familiar friends in new ways can involve no more than observing the last- (instead of the first-) quarter moon. True, both are half-moons that display identical features situated in the same locations, thanks to the moon's lack of apparent rotation. But the illumination is wholly changed, and that makes all the difference. The first-quarter moon is lit up from the right, the last-quarter from the left, and this imparts entirely new shadowing to mountains and craters, the way it would be if you had forever seen your house in the noonday sun and now suddenly glimpsed it at sunset for the first time.

The crescent moon also responds to a fresh perspective. Does

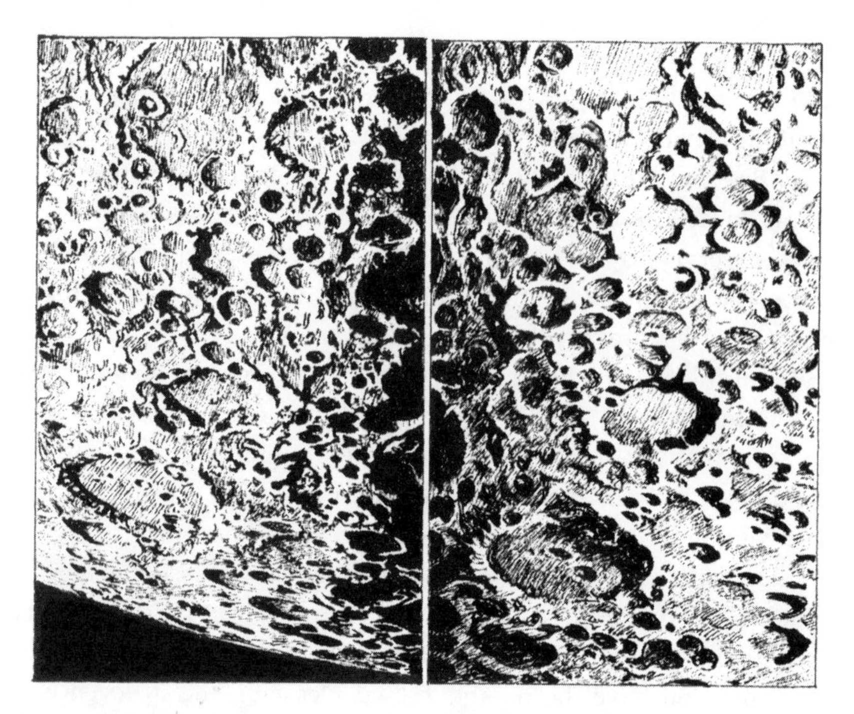

Images of the first-quarter moon (right) *are far more common than those of the third-quarter* (left) *because the latter phase doesn't rise until midnight or later. It's a shame, because we can learn much about the terrain by studying the differently cast shadows, such as those in and around Clavius, the large crater near the bottom.*

it look like a bland smile, a flat pasteup in the sky? See it instead three-dimensionally, where the crescent's inner curve delineates the sunlight/shadow line of a sphere. Your perspective shifts from flat to fabulous. One viewpoint is ordinary; the other conveys an exciting, dimensional *presence* of a globe magically suspended before you—a ball mysteriously hanging in space.

The very way we *can* perceive the universe is new. Our unique nervous systems really just arrived here. After all, 99.9 percent of the species of creatures that ever lived on Earth were extinct by the time humans appeared on the scene. Each undoubtedly had its own biases, its unique ways of apprehending its surroundings. And even within the context of human society, everyday reality changes quickly in our technological age.

At the beginning of the twentieth century, for example, automobiles were touted as a way to stop pollution! Horses were so prevalent (over one hundred thousand in New York City alone), and so much manure accrued (two dozen pounds a day from each horse), that an offensive odor dominated the air. Countless flies buzzed everywhere, spreading a cornucopia of disease. Factor in the noise pollution from the din of horseshoes on cobblestones, and the challenge of avoiding filth when crossing the street, and you see why cars were hailed as ecological saviors: cheaper, cleaner, quieter, and faster.

If cars were once paradigms of environmental health, how many of today's sacred chestnuts will fall away as the new century advances? Or, put another way, how many aspects of life today could we dependably expect to endure for the next century or two? Taxes. Teenage rebellion. Mosquitoes, of course: They've been virtually unchanged for 200 million years, so they'll easily make it through the next few hundred.

Even nature's sky show avoids boredom and repetition. Nothing celestial except for the sun and moon will stay the same. Above the world's light-polluted population centers, stars are already washed out; fewer than one hundred dimly shine in most areas. This trend will certainly continue until only the dozen brightest stars manage to poke through the eerie fluorescent canopy.

The North Star, after becoming more precisely fixed upon true north until the year 2112, will slip away, leaving us with no polestar for several millennia.

Sunset hues, meanwhile, can be expected to continue growing more vivid, their red tones more intense, thanks to airborne particulates. Here, at least, we have an improvement of sorts, at a price.

The moon will keep its same frozen face pointed our way; its rotation has now been locked in place forever. And since the limit of human visual resolution at the moon's distance is about 210 miles, and no human structures are ever likely to be that large, the moon's appearance will not change (as seen from Earth) even if we eventually set up huge colonies there.

The moon's inch-per-year departure from Earth can have no discernable effect for a million years. Nor will the tides' strong fingers alter their grip on the oceans. Clearly, the classical immutables, the stuff of poetry, the grist for the promises of sweethearts will remain even if some incoming asteroid obliterates or remakes our biosphere.

As the armada of colossal new telescopes divulges its harvest, we may mistakenly ascribe newness to previously unknown phenomena. People have always assumed that what lay within view represented most if not all of the cosmos; they have always been shaken by the necessity to take quantum leaps of perspective. Nevertheless, underlying our minds' leaps from petty restlessness to enchantment with the universe's inventiveness, the innate essence of the cosmos always awaits, and it sparkles. Ideas and knowledge

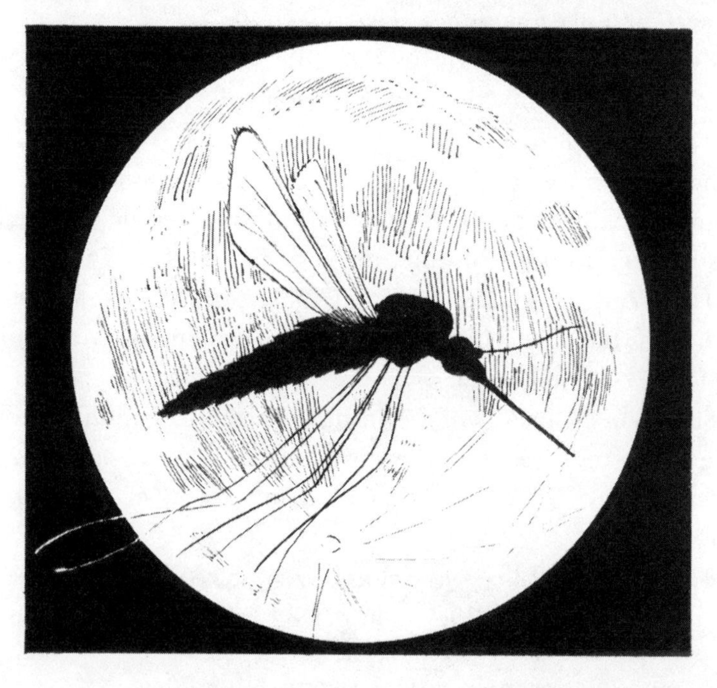

change, but this underlying, playful mystery always invites exploration, ever ready to leap out like a startled deer.

It is very difficult to maintain an innocent perspective in a society of billboards and clichés, where cynicism comes not so much from deliberate skepticism as from overexposure to news media, commercial interests, and even the dualistic nature of language. One effective trick is to abandon the long-held habit of forever deciding what is perfect and what could be improved. To meteorologists, a "perfect" storm is the worst a storm can be: Is a gentle snowfall then really not perfect? Is a meteor shattering chaotically as it tumbles across the sky perfect or imperfect? Is a segment of a rainbow, or a broken wisp of cloud? (Why, then, should we let people's irrational behavior irk us any more than any other natural phenomenon?) Joy's absence seems to come from embracing half measures, for only with a steady focus and full attention is it possible to see *all* as perfect.

Our choice. I'd rather join the infant who gazes at the moon with the rapturous smile of the simpleton. For, in the heavens and on Earth, for observers of the sky and observers of the human pageant, the ultimate reward is knowing: Each just as it is.

Strange Moon

I put a spell on you! says the moon, and it's still there after 4 billion years. Our enchantment with the moon runs deeper even than the passion of poets and the vows of sweethearts and the demented howling of wolves. It permeates our legend and lore, influences weather and behavior, sways billions of tax dollars.

All because we got clobbered by a stray planet (according to the most popular theory), causing a giant piece of molten Earth to blow off into space. Shuddering and rippling itself into a sphere, the way falling water coalesces into droplets, the newborn moon was very close to us at first, occupying a large and dramatic piece of our sky. But it never achieved the critical seven-mile-per-second speed that would have allowed it to escape from Earth completely. If it had, we would have joined Mercury and Venus as lonely moonless worlds; instead, it settled into an elliptical orbit, where it's been slowly spiraling away ever since at the rate of an inch a year.

This Siamese-twin separation left each remnant of the collision, Earth and moon, with permanent withdrawal symptoms. Both for-

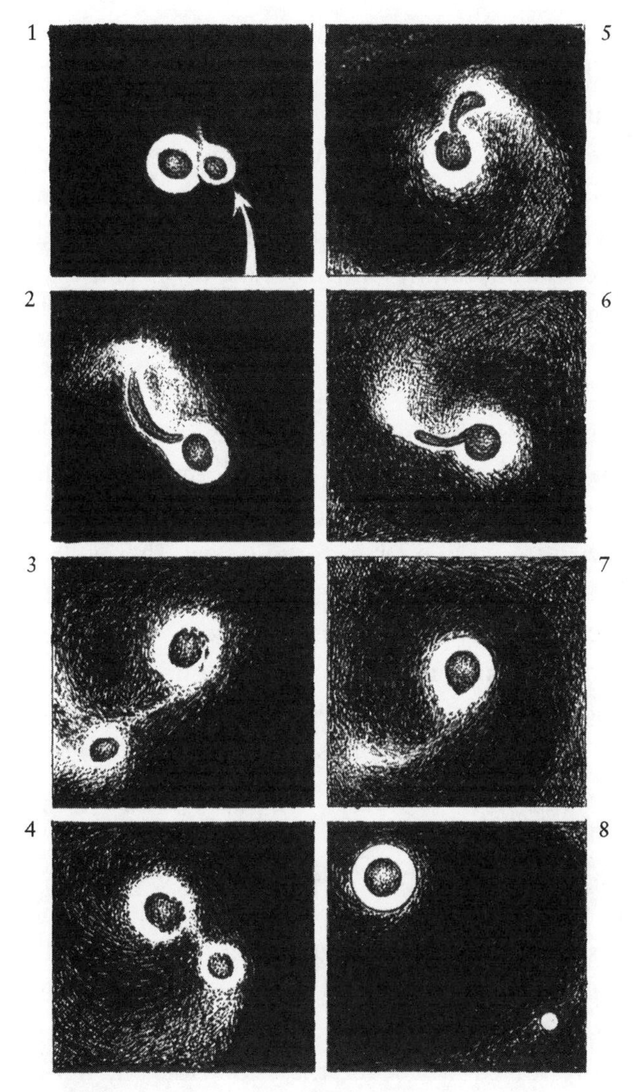

The formation of the moon.

(1) A stray planet smacks the larger Earth a glancing blow. Both have iron cores (dark) surrounded by mantle material (light).

(2, 3) The stray loses some of its core to Earth before rebounding.

(4) The stray pulls itself together but can't escape Earth's gravity and is pulled back in.

(5, 6) The stray crashes again, and its core sinks and merges with Earth's core.

(7, 8) The lighter-weight mantle material scattered by the impact orbits Earth and gradually coalesces to become our moon.

ever feel immense tidal tugs from the other. The larger chunk, this planet, still experiences these as a daily surface deformation of about eight inches: The very ground rises toward and falls away from the moon as we rotate. Additionally, there's the oceanic rise and fall that averages three feet worldwide. And there's a newly discovered atmospheric tide that makes the weather slightly rainier and cloudier during full moon than most other phases. Thus a full moon is somewhat more likely to be hidden behind clouds than, say, a half-moon. The effects go on and on.

Indeed, a few years ago researchers found that the lower few miles of our atmosphere heat up by several hundredths of a degree at the time of full moon. Apparently the moon, despite being as unreflective as a dark country road, both bounces sunlight to us and radiates infrared like a bathroom heater, thermally influencing our world.

The fact that the human menstrual cycle (and that of the opossum but no other animal) approximates (but does *not* match) the lunar synodic period (full moon to full moon) of twenty-nine and a half days bespeaks either a mild coincidence or an earlier forest/jungle era when the darkness of night was linked more closely than now to biological rhythms.

Even educated people (such as maternity-ward nurses) continue to claim a strong connection between birth and lunar phases, contradicting statistics that show human births to be totally independent of the moon's cycle of phases. Apparently this is another case where two obvious visual events (full moon and birth) are noticed and mentally linked whenever they coincide but never unlinked when they do not happen concurrently. Over time the connective assumption becomes reinforced without the person being aware of the process.

The full moon is also, strangely, one of the few *perfectly* round objects encountered in the natural world. Its equatorial bulge of just 4 miles in a diameter that is 2,160 miles across is utterly imperceptible. Because the full moon isn't shaded or a bit extrabright at its center, which would convey the impression of dimensionality, of being a floating sphere, it instead has the look of a flat disk, a dinner plate. Accustomed as we are to it, it's still a curious sight.

But if the moon has influenced us, the reverse is eighty-one times truer: We affect the moon enormously. That little piece of escaped terrestrial real estate, eighty-one times less massive than the larger chunk called Earth, suffers a permanent tug from our world that makes terrestrial tides seem piddling by comparison. That's serious business for the moon, since it possesses a density only a bit more than half of Earth's because of its sandy, silicate composition (probably caused by having been carved from Earth's mantle rather than our metal-rich core). Less able to resist gravitational interference, its side facing Earth (the one that always suffers the greatest tug) was continually braked by Earth's unseen tidal fingers until its rotation froze like hardening cement. This is why the same unchanging face of the moon always appears before us like a stuck movie projector.

We rarely give much thought to the fact that the pattern of dark blotches on the moon (named *maria* by early observers, who thought they were seas) never varies even though it orbits us, rotates, and floats in opposite sides of the sky—all of which should let us see both of its hemispheres. But no. The moon does a bizarre two-step, designed so that one side alone faces our curious eyes. It would be as if we always saw the same side of a horse and rider regardless of whether they were on the near or far side of the racetrack.

You'd think that familiar pattern of moon blotches, the one that makes up the "man in the moon," would be well known to everyone, since, after all, we've all seen it countless times throughout our lives. Not so. We may recognize the features of a few hundred movie stars, but not the face of the moon, our lifelong nocturnal friend. That's why filmmakers can get away with faked, reversed, or upside-down moons. Nobody notices or cares. (We're routinely oblivious about things that don't greatly matter. Can you recall which way you turn a doorknob—clockwise or its opposite—in order to enter a room?)

The moon's shortage of inherent character is symbolized by its total lack of color. The moon is the grayest object in the known universe; visiting astronauts wasted color film on its uniformly monochrome surface. Also, like the ragged hermit who turns out to be a millionaire, the moon differs from its appearance. Binoculars or a small telescope reveal a harsh surface of pointy mountains; in reality it is a land of gently rolling hills. It looks hard and rocky, but astronauts found its landscape uniformly covered by deep and incredibly fine powder, like confectioner's sugar. The austere lunar terrain has been dusted over with a baby-powder softness.

The full moon gives off just 1/450,000 the light of the sun. And while that may seem plenty bright enough to come to the aid of prowlers or lovers on a nocturnal stroll, the moon makes the top-ten list of darkest surfaces in the solar system. Its overall reflectivity is just 10 percent, the same as asphalt. If some crazed developer paved the entire moon—turned it into one huge blacktop parking lot—it would appear no darker than it already does!

Put another way, the Earth as seen in the lunar sky is some sixty times brighter than the full moon looks from here. Translation: If the moon had an Earth-like atmosphere, from there the full Earth would be bright enough to produce a bright blue sky even at night.

There's mystery to the full moon's brightness. One might imagine that a full moon would appear twice as bright as a half-moon. Instead, it's ten times more brilliant. The moon reaches half its full-moon luminosity just two days before the full phase and then quite suddenly finishes its explosive brightening.

The traditional explanation has been an effect called *shadow hiding*, in which the innumerable little shadows cast by lunar pebbles and sand grains when the sun is low in the lunar sky all disappear when the sun is overhead—as it is at full moon. A second effect is the simple bouncing of light (in this case sunlight) straight back at the source, the way a movie screen does; we'd see this light only when the moon is full, since that's when the sun is directly behind us.

Then, in 1995, a new theory took over: *cohesive backscatter.* This explanation for the full moon's brilliance says that sunlight passes through and bounces off internal surfaces in transparent sand particles, producing an additive effect. But, in 1997, the visiting Clementine spacecraft failed to detect the infrared brightening that would have accompanied this effect. So, it's back to good old, simple shadow hiding to explain the sudden and dramatic increase in brightness just before full moon.

But such controversy and confusion merely supply intrigue, rather than any practical impetus or scientific justification for our lunar obsession or the tax dollars thrown its way.

Our relationship with the moon, like our connection with many of our relatives, has always worked best when we appreciated it from a distance. Its 500-degree temperature extreme from day to night, utter lack of air and water, and paucity of valuable minerals or resources have made it thoroughly unappealing as any kind of semipermanent home.

But just when we had totally given up all thoughts of subjecting any additional humans to that wasteland, a momentous and utterly unexpected 1996 discovery that was confirmed in 1998 changed everything. Water! In large depressions on the moon's north and south poles, forever shaded from the sun like some vampire lair, lie lakes of ice.

These frozen oases will make all the difference. Instead of needing to haul every drop of water at hellacious cost (at eight pounds per gallon, water is a heavy substance extremely expensive to transport), we can now count on a supply awaiting us at the lunar poles that should be easily extractable from the soil.

Instantly, the discovery changed the course of human ambitions and possibilities. Instead of a worthless, uninteresting, dry wasteland, the moon is now a worthless, uninteresting wasteland with two freezers full of ice. If we ever want to make it a way-stop, a

Looking at Earth over the south pole of the moon, we see the depression and deep craters that provide perpetual shade for water ice deposits. (There is actually less sunlight reaching the pole than is shown here.)

jumping off point to far places, it has just gone—bingo—from impossible to possible. It could even serve the same role as the old wooden railroad water towers that serviced steam locomotives, as a watering station for interplanetary shuttles.

Despite the hype of the mass media, not many people really believed we'd be anxious to return when the final *Apollo 17* astronauts left the moon in December 1972. And the quarter-century hiatus before the idea resurfaced in 1998 only proved how prohibitively expensive and unappealing such a return would be. But of the ancient elements of air, fire, earth, and water, only the latter really matters; ice on the moon may prove to be the most significant discovery of the twentieth century so far as human exploration of the universe is concerned. Those ebony sun-shielded craters have changed everything, their import not to be fully gleaned until the passing decades turn the hereafter into history.

If we do go back, maybe we can remove or modify the plaques

that the six Apollo landing teams left behind between 1969 and 1972. All have in common a single signature and the accompanying name of the then American president, Richard Nixon, put there at his insistence.

As things now stand, that name will survive a half-billion years in the moon's airless environment, long after every trace of human existence has vanished. Visiting aliens who find no sign of humans on Earth would know that we once were here if they visited the moon. Their experts would surely wonder what could be the glory and grandeur of this single person, that his name would be exalted and memorialized beyond every other person who ever lived. Obviously, we must remedy this. The best course might be to clarify the plaques by adding one of Mr. Nixon's famous quotes. My first choice would be "I am not a crook."

Bohdan's Bursters

A small army of brilliant men and women spend their lives probing the universe—and nearly all of them are utterly unknown to the public. A handful, like Britain's theorist Stephen Hawking, tend to get most of the publicity. A peek into their gifted minds can boost us to the dizzying edges of space-time, but chronicling their labors would entail a small library unto itself. So my solution is to profile the life and work of one, serving to represent a cross section of the amazing cosmic enterprise we have embarked upon. It staggers the mind.

The man I'm spotlighting serves as an excellent example of how the strangest phenomena surrender to our most luminous intellects. The choice is easy. First, because I came to know this witty and gracious researcher, Bohdan Paczynski (say Pa-CHIN-ski), when interviewing him for an *Astronomy* magazine article in the summer of 1997. Second, because his work is particularly fascinating, as it touches on not one, but two of the most amazing phenomena in the universe.

I quickly learned that Paczynski loves things that wink on and off like holiday lights. I used to observe variable stars and am still intrigued by light/sound/animation gizmos; this was enough to establish an immediate affinity. His simple obsession has led him to become one of the most respected voices in astrophysics. Paczynski proved to be not merely ahead of the times; he also happened to be right—a fact that propelled him to the edge of the universe and the forefront of today's hottest and most bizarre astrophysical problems.

Born in Poland, Paczynski, through his perceptive early papers, earned an invitation to become a research assistant at Lick Observatory in California, where he proved himself outstanding in his field as well as somewhat accident-prone. For example, he managed to doze off one night and fall from the observing platform. (No harm done, other than a ten-minute gap in his observation program.)

While at Lick, he bought his first car and learned to drive on the tortuous, video-arcade-like Mount Hamilton Road overlooking San Francisco Bay (seven hundred turns along just twenty-four miles). Told that a driver can make just one mistake, and shown car wreckage down the canyon to emphasize it, Paczynski proved the experts wrong.

But barely. He went off the road and still lived to publish a dozen papers derived from his Lick visit. His survival involved more than simple Lick luck; his continued existence was a direct consequence of the very low suspension of that era's American autos. Two wheels went off the road and hung suspended far out in the air over the canyon, while the car's bottom caught at the road's edge.

Watching Paczynski check his E-mail from his cluttered Princeton office and type intricate formulas clickety-clack with two fingers stabbing the keyboard like woodpeckers, I wondered where the leap occurs. At what point does that mysterious illuminating *something* allow the brightest among us to vault across the chasm that connects the human mind with the soul of space-time?

Back in 1986, when orbiting satellites had already been telling us for nearly twenty years that bursts of the highest type of energy—gamma rays—were popping daily like party balloons in all parts of the sky, it was anybody's guess what they were and how they were produced. This was new stuff, a phenomenon never before experienced. It was the kind of novel now-see-this that comes along only every few decades, as quasars did in the sixties. It was the sort of event astronomers dream about, like their caught-in-a-black-hole nightmares after too much pasta fazool before bedtime. And yet this was real: Like flashbulbs igniting at random in a crowded stadium, the bursts lasted from a few thousandths of a second to several minutes. One even managed to persist for about the length of a feature film.

There were two ways to look at this bizarre affair. The first (which virtually everyone thought at the time) was that they lay within a few hundred light-years of us: The explosions were happening in our galaxy. Maybe they were fireworks produced by starquakes on nearby neutron stars. Later, most astrophysicists pushed them a bit farther out—to the vast halo of invisible material in which the Milky Way sits like a ship in a bottle. Placing them relatively nearby allowed their energy output to be merely astoundingly high instead of off the chart.

Paczynski instead argued that they hovered near the edges of the observable universe and that their distribution provided the critical distance clue. If they were Milky Way objects, we'd primarily find them scattered along the Milky Way, peppered amid the plane that rings our sky like a glowing highway. If, instead, they were part of our galaxy's spherical halo, they'd still not show a uniform distribution because of the inconvenient fact that Earth is not centered in this city of suns. Since our planet's little plot of celestial real estate sits in a spiral arm that's 28,000 light-years from galactic downtown, the burster distribution should follow the pattern of other halo objects such as globular clusters—that is, skewed in one direction. But this, too, isn't the case. Instead, they exhibit a perfectly random distribution. And unless you accepted what Paczynski saw as implausible rationalizations, this meant they faithfully followed the configuration of faraway galaxy clusters and therefore lurked at cosmological distances. Anything else made him squirm. He quotes astrophysicist David Schramm: "A theorist is allowed to introduce one tooth fairy into his theory—but *only* one."

To Paczynski, the reluctance of most astrophysicists to accept what he considered to be the obvious, even if it opened up new cans of wormy things, was unsettling. "I'm amazed," he says, "how many ad hoc assumptions some people were willing to make in order to keep bursters in the galaxy. Why bother?"

The enigma was deepened not just by the iffy observations, but by the transient and inconsistent nature of the bursts themselves. Paczynski believed that "any theory that explained all the observations had to be wrong—because some of the observations were wrong!" Even the widely respected Martin Rees of Cambridge University offered Paczynski hundred-to-one odds that bursters were *not* cosmological. Paczynski declined the wager, prompting Rees later to quip, "We were both fools."

Yet with a truly great man's lack of hubris, Paczynski readily confesses that a bias in the data selection process made him base his early conclusions on incorrect evidence. He regards as undeserved the credit given him for properly placing bursters at the edges of the universe: "I was lucky."

While the controversy raged, about three hundred bursters a year continued to pop, and the Compton Gamma Ray Observatory with its octet of detectors (called BATSE) kept catching them in the act.

The problem was that BATSE couldn't pinpoint their locations accurately enough for truly large telescopes (which have tiny, squinty, pinhole-like eyesights) to be focused exactly on target. So, as the years dragged on, not a single burster could be associated with a celestial object at any other wavelength. Not one optical counterpart was found where bursters had gone off. Nor was there anything in radio wavelengths or in the ultraviolet. Either these bizarre creatures gave off *only* gamma rays (which is as odd as the sun deciding to emit green light and nothing else), or else we kept failing to turn a powerful enough instrument to the right spot at the right time.

The right time. That was the key. If these objects were so transient that they'd usually vanish faster than you could say "peculiar," maybe *something else* would linger, like a Cheshire cat's smile, in the spot where they once stood.

Enter the new Italian/Dutch satellite with the appropriately cryptic name BeppoSAX (an acronym even more obtuse than most, since it's derived from non-English words). The insuperable problem had been time—and BeppoSAX provided it. With the satellite's sensors capable of better pinpointing the bursters and the ability to hop to it within hours instead of days, ground-based observatories could now swing their instruments and see if anything was there.

On February 28, 1997, it finally happened: BeppoSAX delivered its first quick and accurate position for a burster. When X-ray and other instruments checked it out, this burster (and subsequent ones on April 2 and May 8) displayed afterglows at various other wavelengths. At last, the rest of astronomy's arsenal could pounce on the burster bandwagon.

The fifteen-second-long burster on May 8 really did the trick, since BeppoSAX provided a position accurate enough for the Keck Telescope in Hawaii, the world's largest, to swing into action three nights later. It found a twentieth-magnitude point source embedded in a galaxylike smudge at the burster's location. The dot was a million times fainter than the dimmest thing visible to the naked eye—but that was bright enough to take the first-ever burster spectrum at optical wavelengths. It revealed an intervening gas cloud so red-shifted that it had to lie several billion light-years from Earth. Which meant that the source light, the burster, lurked farther away still. The astronomical community was finally convinced: This burster sat 8 billion light-years away, more than halfway to the edge of the universe.

For bursters to appear so bright at such distances, their energy output has to be mind-boggling. Consider: In a single second, a burster emits more power than the sun has given off in all the 5 billion years since its birth. Such unimaginable energy, channeled into a fraction of a second, is like some cosmic photographer saying "Watch the birdie" and then setting off a flash that makes a supernova seem by comparison like a car backfiring, an energy pulse stretching clean across the universe.

Nobody knows what can produce that kind of power. Of course, this doesn't stop theorists from guessing. Paczynski favors the notion that an ultrapowerful magnetic field enhances, focuses, and channels some kind of energy release so that it's aimed our way. If that energy is pointed toward us like a laser beam, then less power is emitted than if such energy were flying off every which way. In that case, explanations that involve collisions between neutron stars can be made to work. We're spared the possible hassle of coming up with a new set of physical laws.

England's Martin Rees may be on the right track in suggesting that some kind of fireball (maybe from an ultrastrong supernova, called a *hypernova*) produces a burst when its shock wave interacts with surrounding gas, and ultrastrong magnetic fields extract and channel some of this energy like a kind of cannon. In this *fireball model,* a small amount of mass, perhaps just a millionth of our sun's substance, may be expelled at a speed approaching that of light. Interacting with a magnetic field an order of magnitude stronger than any yet seen, a quintillion (10^{15}) gauss—which definitely would have no trouble sticking to your refrigerator door—it could produce a burst of gamma rays directed along its polar axis. The afterglow would come from the material interacting with its interstellar medium. Voilà. Explained. Sort of. To Paczynski, it's a beautiful example of how things can work out when nature is kind enough.

While this overall picture seems as attractive and reasonable as the notion of dismantling Congress, the specifics of what bursters really are remain to be determined. And there it sits.

But if gamma-ray bursters are weird, then Paczynski's other

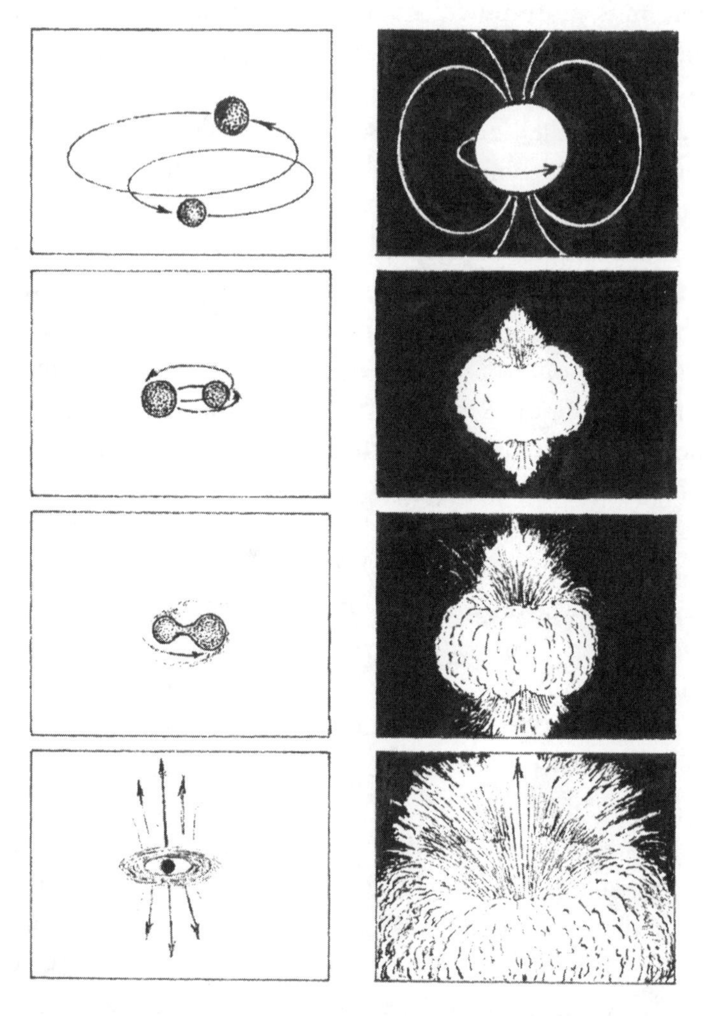

How to explain gamma-ray bursters, the most energetic events in the universe?

One theory (left) *postulates that a bound pair of neutron stars* (top left) *gradually fall together to produce a black hole, at which point its accretion disk ejects matter at near light-speed in polar directions.*

A different theory (right) *suggests that a rapidly spinning (and therefore highly magnetized) giant star explodes. The material directly over the supernova's pole is least impeded by the magnetic field and escapes at near light-speed, and we happen to be aligned with one of the poles.*

claim to fame—his favorite project and personal brainchild—is weirdness squared. Imagine: You start with Einstein's amply proven fact that space can warp. Then, as suggested by Paczynski a decade ago, you use that curved space to find the hidden dark matter in our galaxy. Simple.

Within a few years, to Paczynski's surprise, newly evolved technology allowed researchers to do exactly that, and teams from four countries, including a group of Polish Paczynski collaborators, got to work. The idea was to ignore truly massive objects that really do a number on their surrounding space. (Although, granted, such phenomena are spectacular; an image from the 1996 *Astrophysical Journal* shows a blue galaxy appearing in four different places in the sky because its light had to pass through the curved space surrounding a foreground galaxy cluster.)

While such fun-house distortions are astounding cases of gravitational lensing (with each image separated by the apparent width of Jupiter), what Paczynski had in mind was *microlensing*. That is, he suggested surveying *tens of millions* of stars in a crowded region. His project, called OGLE, chose the direction of our galactic center; a Berkeley-based team called MACHO picked our companion galaxy, the Large Magellanic Cloud. If he repeatedly checked the light of each of these millions of suns, Paczynski reasoned, sooner or later a few of them had to pass behind an unseen foreground object and have their light distorted. If the alignment was exact, there'd be a glowing halo, the bizarre Einstein Ring. A less precise lineup would force the background star into a series of eerie, twisted crescents surrounding the invisible foreground whatever-it-was.

Paczynski knew there'd be no way to directly see those lensed star fragments, spaced apart by just a thousandth of an arc second. But as multiple images magically appeared, a characteristic brightness change *should* be detectable by simple photometry.

Okay, not so simple. Daily monitoring tens of millions of stars isn't the kind of thing you'd want to attempt in your spare time. But CCDs (charge-coupled devices) and the new computer programs do it automatically. Not surprisingly, the first thing such

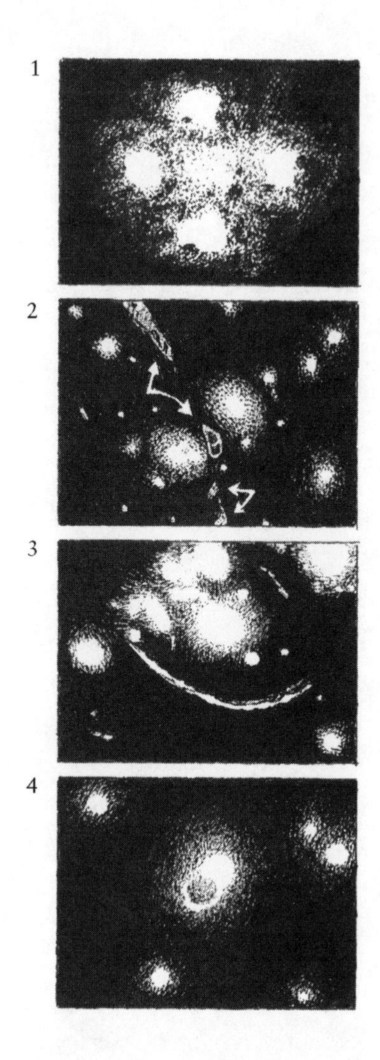

Cosmic lensing:

(1) Einstein's Cross consists of a distant galaxy (center) *that bends the light from a quasar some twenty times more distant into four images.*

(2) Multiple images (arrowed) *of a distant galaxy, distorted by a foreground galaxy cluster.*

(3) Here the light of a distant galaxy is lensed into an arc of light by the nearer compact galaxy cluster.

(4) Paczynski hoped we could observe the lensing of stars in our own galaxy by intervening dark objects. Such an object might look like this if highly magnified. An actual, complete "Einstein ring" was finally observed in 1998.

intense photometric surveys found was a zillion new but mostly ordinary variable stars, which made Paczynski perfectly happy. Remember, he loves things that wink and blink. He points out that serious amateurs with good equipment can discover countless variables as well. Once you go below eighteenth magnitude, it's all new territory.

In the fall of 1993, the first actual lenses were discovered almost simultaneously by three groups of researchers.

What are these unseen bodies? With one data point taken per night, the technique was insensitive to both very short and very long lensing events. Since low-mass objects produce shorter events and because the procedure would best detect phenomena lingering from ten to one hundred days, the project was expected to uncover microlensing caused by objects weighing about a quarter of what the sun does. Since that's exactly the mass of most ordinary stars, you can easily follow Paczynski's conclusion that while the study is still ongoing and preliminary, the 250 lensing events detected between 1993 and 1997 were probably not strange new exotic material, but invisibly faint but common foreground stars.

You make billions of measurements, and all you find are simple table-salt-variety red dwarfs, the street dogs of the galaxy. A bummer? A letdown?

No way. If we still don't know what most of the universe is made of, there are other fish being fried here. First, the successful results prove that microlensing really works. We're only at the dawn of its discoveries. Second, the number of lensing events have shown more material than expected in one particular direction, evidence that we really do live in a barred galaxy and that the central bar is more or less pointed in our direction. Third, the search for dark matter is so fashionable, it helps keep these microlensing projects funded.

From his home base at Princeton University, Paczynski continues to inspire the search for new things that blink and change, enabling the rest of us to follow him and his gifted peers—past and present—to new frontiers of galactic knowledge.

In many there is the one: *E pluribus unum*—one person's dedication exemplifies our species' obsession with the cosmos. The totality of astronomical activity is overwhelming, confusing; the pursuits of a single one of us suddenly delivers one more gift of understanding and appreciation.

Writing here about Bohdan Paczynski, a man unknown to the general public, is my tribute to him as a dedicated scientist—a symbolic spokesperson for so many other unheralded geniuses worthy of inclusion in the pantheon of astronomy's greats, those who quietly search for missing pieces to the grand puzzle by reaching for the stars.

Top Hits of the Cosmic Charts

Are we exploring the universe because of some unique cerebral hardwiring? Do our underused, oversized brains have some kind of neural feedback loop that fosters this expensive craving to build colossal observatories and orbiting telescopes—all so we can fill their excess capacity with endless new questions?

I think part of astronomy's appeal may have a much simpler origin, and it isn't intellectual at all: Every once in a while, lucky people are treated to an enrapturing experience under the night sky and simply want to repeat it. This helps explain why stargazers, offered a galactic menu, point their telescopes toward the same handful of objects time and again. Some people marvel that I can so quickly locate Saturn in the confusing, twinkling quilt of the heavens; not remarkable in the least, considering I've already performed the trick a thousand times. The only skill I'm *really* demonstrating is a mastery of inertia.

It's part of the frailty of pleasure addiction, which also propels us to do risky things like marrying on impulse or flash-microwaving a pint of hard-frozen ice cream so that we can gobble it down in a minute or two. Some nighttime activities can be just as pleasurably hazardous; collectively they've given us enough remarkable nocturnal moments to keep us hooked forever.

It all may have started naturally, with early memories of sizzling meteors and mind-paralyzing Milky Way vistas. But now, as adults, we're jaded: Only the splashiest night show will earn our Oscar.

What's needed is a guidebook for the celestial hedonist. One that's focused on the greatest spectacles of the sky, the spellbinders that pull us back to this cosmic theater time and again, for life. College astronomy courses bestow thorough introductions, popular books offer specialized celestial pursuits, yet a gap remains: None of these addresses the impatient urge to cut to the chase and experience the ultimate thrill.

We realize the need for such condensed excitement whenever we're trapped with a fanatic, someone involved with a particular branch of science—geology, say—who goes on and on about a subject like pumice. We may grit our teeth, but we enjoy the person's enthusiasm. If only it could be distilled to the point where we hear or see the most fascinating things and are spared the endless particulars. (Like my feelings about opera: Love the great overtures and arias, hate the rest.)

So, too, with the sky: Most people want only the best, the most absolutely stunning celestial sights. Not merely beautiful things, like Orion's return each autumn, or the view of an edgewise galaxy through a large telescope, but a step beyond to the profoundly mind-blowing.

If only heart-thumpers could be summoned at will! Early on, I fine-tuned myself to people's reactions when viewing the most dramatic celestial phenomena so I could anticipate and savor the vicarious thrill of experiencing them again through fresh eyes. I never tire of that. The reactions are consistent and span all cultures. Tibetan rimpoches and lamas who have visited Overlook observatory (descending from the Buddhist monastery atop this same mountain) reacted to a spherical city of suns named M13 in the same awestruck manner as a visiting congressman. On one particularly notable occasion, while vacationing on the beautiful island of Mooréa, near Tahiti, I was surprised to encounter my hotel's cook, who, like me, was setting up to observe on an ink-black beach. He had a fine eight-inch telescope and, without any suggestions from me, turned it to the very same sights I would have chosen. It seems the astronomy pleasure-instinct, as well as the celestial objects that satisfy it, is universal.

For the unaided eye, truly memorable apparitions need to be more powerful and, to be fully appreciated, more unusual—almost everyone devalues the commonplace. A beautiful sunset's maelstrom of original patterns would be exalted and written about for generations if it appeared just once per millennium. But as things stand, most people don't bother looking out the window to see it.

What follows is a subjective list, in order of awesomeness, based on thirty-five years' observations, of the greatest spectacles the sky can offer—the inspiration for a living love affair with astronomy.

The premier sky spectacle has to be, no contest, the total solar eclipse. It's caused by the astonishing coincidence that the sun and moon, the only disks in Earth's sky, both appear exactly the same size! (The sun is four hundred times larger, but also lies four hundred times farther away.)

This rare event shouldn't be confused with partial eclipses, which are viewed by most people every few years. Totality is as different from a partial eclipse as night from day. And, as if to ensure that the greatest spectacles also remain the most unusual, totality happens only once every 360 years, on average, for any given earthly site. (And if it's cloudy, you have to wait *another* 360 years!) There hasn't been any totality happening anywhere in the mainland United States since 1979, and there won't be another until August 21, 2017. Which means, unfortunately, that our nation is in the midst of an unprecedented thirty-eight-year eclipse drought.

Bottom line: You have to travel to see them. Since totality occurs *somewhere* in the world nearly once a year, it necessitates a pilgrimage—usually to a place about as accessible as Mongolia.

That it's amply worth the journey is borne out by the emotional experience of all who attend. Half of those who witness totality groan, moan, shout, and behave strangely. It's not just the chickens who act up when the sun, moon, and Earth form a perfectly straight line in space. Something happens: I'm convinced that a blind person could easily sense the primal intensity of the event. It's about as close as you can get to letting your nervous system regress all the way back to the Cro-Magnon.

The sun's strange, glowing corona, or atmosphere, offers a unique, ultrahot, million-degree-plus radiance that dramatically leaps across the sky at this occasion—but at no other. Because its curiously shaped inner segments are bright but the outer regions are faint and lacy, photography fails to capture the entire coronal presence, yet the human eye, with its sensitivity to a large brightness range, manages it with ease. This is why a total eclipse looks and feels wholly different from its photographed or televised image.

As if all this weren't enough, various ancillary phenomena materialize. One especially noteworthy phenomenon: *shadow bands.* These are shimmering streams of curvy black lines that usually ripple across every white surface, such as a sandy beach, during the minute before and after totality. An astonishing fact is that they cannot be photographed! Dramatic to the naked eye, they're absent from a developed snapshot. Even if you didn't know this, shadow bands are still eerie enough to raise the hair on your neck.

A total eclipse is so totally irresistible that having once experienced it, people have been known to take second jobs or mortgages, do whatever they must, to get to the next one.

Amazing sky spectacle number two is the even rarer meteor storm. A hundred "shooting stars" each *second* can radiate from one point in the heavens and continue for more than an hour. Looking like a wildly sputtering short circuit from another dimension, the event generates bewilderment and fear more than wonder and joy.

One woman I know, upon unexpectedly observing the 1966 meteor storm from a train crossing the Texas countryside, was sure that the end of the world had arrived. What else could it be? Looking around the compartment, she saw that everyone else was sleeping, unaware that the heavens were exploding furiously. She anguished over what to do: Should she awaken these strangers? What was the correct protocol? Do you wake up people for the end of the world or let them sleep through it? (Then a conductor came by, and together they watched the astounding phenomenon until dawn brightened the sky.)

Meteor storms are caused by dense concentrations of mostly apple seed–sized rocks and ices that collide with Earth. The last such event occurred on November 17, 1966, after an absence of almost a century. The next possible display is due on November 17, 1999, a fine celebration of the century's end, although there are at least even odds that it won't happen at all this time around.

Phenomenon number three is probably a great comet. Two crossed our skies in consecutive years, 1996 and 1997, when comets Hyakutake and Hale-Bopp graced the heavens (both in the month of March, just like the previous noteworthy comets, so-so Halley in 1986 and gorgeous Comet West in 1976). From dark skies, Hyakutake was certainly the more spectacular of the two, although far more people witnessed Hale-Bopp because of its greater brightness. Neither was remotely as dramatic as some of the finest historical comets: Hyakutake's tail could have used more brilliance, while Hale-Bopp's fine tail was only a tenth as long as the truly amazing comets of history such as the Great Daylight Comet of 1910 or Comet West of 1976.

A great comet is usually discovered only a couple of months before it becomes spectacular and then stretches motionless across the heavens, sometimes spanning half the sky. Glowing on its own, its presence ineffably awe-inspiring, it's a sight that must be seen to be believed. Nobody can predict when the next one will arrive.

In my book, awesome apparition number four is a grand display of the northern lights. The aurora borealis is customarily linked to penguins or polar bears. But the waving curtains of riotous greens and reds are not confined to distant frozen landscapes. The eerie spectacle often appears over dark skies throughout the northern half of the United States.

In upstate New York, where my observatory is based, I have seen displays of the northern lights that shimmered crazily across the entire heavens, producing new scenes each fraction of a second. These alchemists of surprise can exhibit blotches, rays, bands, arcs, or curtains, or any mix-'n'-match combination. They can appear to float supernaturally motionless, or pulsate leisurely, or flicker

Even small telescopes were able to detect dramatic shells of dust and gas surrounding the nucleus of Comet Hale-Bopp in 1997. The shells are jets of material streaming off the tumbling comet.

rapidly with five entirely new scenes every second, with the observer's spirit seeming to vibrate along as if in chorus. Their silence is otherworldly, and so are occasional reports of crackles or hisses, attributed to the controversial ability of humans to sense the huge electrical charges on the ground beneath them.

Luckily for residents of the United States, the northern lights are usually sections of an extensive glowing halo centered over Earth's northern *magnetic* pole, where compasses point, rather than the geographic pole around which the planet rotates. Thanks to an accident of geography, the magnetic pole lies exactly north of the central United States, some 13 degrees of latitude closer to us than the pole of rotation. At our comparable latitude in Europe or Asia, people are situated 1,800 miles farther from this *auroral oval* and experience far fewer displays.

An aurora is noticeable every month or so from states such as Montana or Minnesota, or even from New England and the Pacific Northwest. Once every year or two, a fabulous, brilliant, sky-spanning event blazes across the heavens, especially during the times when the sun reaches its maximum in its eleven-year cycle of solar storms.

When a solar storm faces us, it may send a blast of electrically charged particles across Earth's magnetic field, generating the electricity that excites electrons in the atoms of our upper atmosphere. This is serious power: Twenty million amps at around 50,000 volts are typically funneled into the regions above the magnetic poles, causing those atoms to glow like fluorescent tubes.

An aurora can appear as an unimpressive greenish patch, or it can grow into a mind-numbing psychedelic light show. As electrical currents are generated on the ground, as if to mimic the sky above, power surges whip through transmission lines. On March 13, 1989, during an intense aurora that was seen all the way to Florida, such sudden currents caused transformers to fail (throwing nearly the entire Canadian province of Ontario into night-long darkness), while garage doors throughout North America kept opening and closing from midnight until dawn.

The last solar minimum occurred from 1995 to 1996; now the sun's storms are on the rise again, a portent of increased northern lights. With a maximum predicted for the year 2001, we're now entering the period when people lucky enough to live in a rural area away from light pollution should acquire the habit of occasionally glancing northward, even when simply coming home and walking from the car to the front door.

Spectacle number five is a *bolide,* or fireball. While auroras offer leisurely spectacle, an exploding meteor is sudden and short-lived. There's no predicting when one will rip through our atmosphere, except to say that they happen once or twice a year above every location. Unfortunately, except for insomniacs and people on the graveyard shift, the night finds most of the countryside snoring. When, without warning, a colorful meteor explodes, it usually benefits only pines and rooftops. As a bolide lights up the heavens, trailed by glowing cascading fragments, it may manifest as a sudden brilliant glow beyond bedroom curtains. Light sleepers may be momentarily aroused, only to shrug it off as "heat lightning."

Great sky apparition number six could easily be a double halo. Meteorological rather than astronomical, it's an amazing sight, straight out of Oz. I've seen only two in my life, the sun surrounded by a pair of brightly colored concentric circles. When it occurs, the inner ring is always the striking (but curiously underobserved) 22-degree (in radius) halo that appears among cirrus clouds several times a month from most locations. This common phenomenon may often go unnoticed because it is such an unexpectedly big apparition. Yet it's absolutely dwarfed by the much fainter and rarer outer ring, which boasts an astounding diameter of 92 degrees, stretching from the horizon to the zenith!

Telescopes can extend the number of great apparitions. But since the preceding half-dozen materialize to the naked eye, without the necessity for gadgets or technological intercession, they carry an especially powerful impact that can sweep even a jaded soul into stunned intimations of infinity.

The price of admission is the habit of occasionally glancing

An exploding meteor can cast shadows as its fragments burn and plummet.

upward. Think about it: Half of these wonders (numbers four through six) occur without warning, a gift reserved only for those who happen to be watching the sky.

I've asked many professional astronomers what brought them to the stars. Most relate some early childhood experience involving one of the above, if not any of the dozens of less powerful but still influential sky spectacles, such as a first peek at the moon through a small telescope. So the cosmos itself seems to create astronomers. We do not consciously decide to explore the heavens. The universe, by dazzling an impressionable youngster with some unexpected spellbinding sight, seems to handpick those who will spend their lives in search of its mysteries and in awe of its glories.

But even those who do not sky-watch professionally received their own almost irresistible invitation when they first glanced upward, to come along for the everlasting ride.

Glossary

ABSOLUTE ZERO: Lowest possible temperature. At absolute zero (−459.67 degrees Fahrenheit; −273.15 degrees Celsius), there is no energy available for transfer to other atoms.

ANDROMEDA GALAXY: Nearest spiral galaxy; also the most distant object generally visible to the naked eye. Also known as M31, Andromeda is a spiral galaxy similar to our own Milky Way but nearly twice as large, lying 2.2 million light-years from Earth and containing about a trillion stars.

ANGULAR DIAMETER: Measurement in degrees of the size an object appears in the sky. For example, the sun and moon each have an angular diameter of about one half of a degree.

ANTIMATTER: Matter with electrical charges reversed. An antimatter atom contains positively charged electrons (positrons) and negatively charged protons. When matter and antimatter meet, they explosively convert all of their mass to energy, releasing energy of 511,000 electron volts in the gamma-ray part of the spectrum.

APPARENT MAGNITUDE: Brightness of an object as seen from Earth.

ARC MINUTE, ARC SECOND: In angular measure, small divisions of a degree. There are 60 minutes in a degree and 60 seconds in a minute (therefore 3,600 seconds in a degree of arc). These angular units should not be confused with units of time. The human eye's resolution is about three arc minutes; anything smaller than this requires optical aid.

ASTERISM: Pattern of stars that forms a familiar figure but which is not a separate constellation. Familiar asterisms include the Big Dipper, the Seven Sisters, and Orion's Belt.

ASTEROID: Also called a minor planet, a small, airless body usually orbiting between Mars and Jupiter in the so-called asteroid belt. Asteroids may represent planets that never formed because they were caught between the tidal tugs of the sun and Jupiter. They range in size from Ceres, 600 miles (965 kilometers) in diameter, to boulders a few feet across. Occasionally, asteroids collide with one another to release dust particles that intersect our own orbit, producing visible meteors.

ASTRONOMICAL UNIT (A.U.): Average distance from Earth to the sun. Distances within the solar system are usually expressed in astronomical units. An A.U. is equal to 92.9 million miles (149.5 million kilometers).

BIG BANG: Most likely explanation for the beginning of the universe. This theory, originated by Aleksandr Friedmann (1888–1925) and Georges Lemaître (1894–1966) in the 1920s, maintains that the universe was created around 15 billion years ago with a primordial fireball containing all space, time, and matter, which began expanding explosively. Since the laws of physics as we know them did not apply before the Big Bang, it is impossible to say what existed "before" the Big Bang. This theory is supported by the presence of cosmic background radiation.

BLACK DWARF: White dwarf that has completely cooled. It is not known whether our universe is old enough for any black dwarfs to have formed.

BLACK HOLE: Compact, crushed object of infinitely high density. The black hole's singularity (containing all its mass) takes up less space than the period at the end of this sentence yet weighs more than the sun. Black holes may form from the remnants of supernovas, whose cores collapse if they weigh more than 3.4 times that of the sun, or they may represent the evolutionary end point of massive suns. Some weigh as much as 3 billion suns, as seems to be the case for the supermassive black hole in the core of the galaxy M87.

BLUE MOON: Astronomically, a second full moon within the same calendar month. This happens every two and a half years, on average.

BROWN DWARF: Object that is almost a star but lacks sufficient mass to produce the high temperature and pressure needed to start and maintain nuclear reactions.

CEPHEID VARIABLE: Class of stars named after Delta Cephei, the first Cepheid variable star discovered. The brightness of these stars changes in a regular rhythm, and the length of time it takes them to oscillate is linked to their absolute brightness. Because of this relationship, astronomers can determine the distance to galaxies where Cepheids have been identified.

COMET: Small body, orbiting the sun, whose volatile material and highly eccentric orbit separates it from an asteroid. As a comet approaches the sun, the latter's heat sublimates some of the comet's ices into a tenuous tail, millions of miles long, which is pushed away from the sun by the solar wind.

CONSTELLATION: Star pattern. Most of eighty-eight constellations recognized today, including the twelve zodiacal constellations, are the same ones that were listed by Ptolemy in 150 B.C., and most represent Greek mythological figures.

CORONA: Outer layer, or atmosphere, of the sun, made of plasma. Although its size varies with the sun's magnetic activity, the corona is generally about 8 million miles (13 million kilometers) thick and has a temperature of roughly 3 million degrees Fahrenheit (1.67 million degrees Celsius). The corona's brightness is equal only to about that of a full moon and so the corona is visible only during total solar eclipses.

COSMIC RAY: Charged particles traveling at nearly the speed of light. Cosmic rays are generally protons (nuclei of hydrogen atoms), alpha particles (helium nuclei), nuclei of other atoms such as beryllium and lithium, and electrons, positrons, and some antiprotons. Their origin is uncertain.

COSMOLOGY: Study of the structure, birth, and evolution of the universe as a whole.

DARK MATTER: Mass that is thought to constitute 90 to 99 percent of the universe. Rotation periods of galaxies, including our own, and the speedy motions of individual galaxies within a galaxy cluster, prove that there is much more material exerting a gravitational force than can be seen. The nature of this material is unknown.

DEEP-SKY OBJECT: Any visible object not belonging to our solar system.

DOBSONIAN TELESCOPE: Inexpensive, basic telescope that swivels like a lazy Susan. They are valued by amateur astronomers for their economy and were popularized by John Dobson, a San Francisco–based amateur astronomer.

DOPPLER EFFECT: Change in the frequency of an object's radiation as it approaches or recedes from an observer. This effect was first studied by Christian Johann Doppler (1803–1853) in 1842.

DOUBLE STAR: Two stars orbiting around a common center of gravity (barycenter). Forty-six percent of all stars are thought to be members of double-star systems, and 39 percent of stars are part of multiple-star systems (with more than two stars), leaving single stars like the sun in the minority. The orbits of double stars, which follow very regular paths, may be as short as a few hours, as in the case of Algol and its unseen companion, or as long as millions of years, as in the case of Proxima Centauri and Alpha Centauri. The closer two stars are, the shorter their orbital period. Some stars, called *contact binaries,* are close enough to actually touch. If double stars pass in front of each other during their orbit, they are called *eclipsing binaries.* Often two stars will start out separately, but one may donate some material to the other. If the secondary star is a white dwarf and enough material is added to it, it may explode into a supernova.

EARTHSHINE: Illumination of the dark portion of the crescent moon by the Earth's reflection of sunlight. Earthshine is our own light, reflected back to our eyes from the moon's surface.

ELECTRON: Along with protons and neutrons, one of the building blocks of atoms. An electron has a charge of 1.602×10^{-19} coulomb and a mass of 9.108×10^{-28} gram. Electrons have a negative charge and are never found in the nucleus of an atom.

ESCAPE VELOCITY: Speed an object must attain to escape from a gravitational field. The Earth's escape velocity is 7 miles (11 kilometers) per second. The escape velocity of an asteroid, which has very little mass, is so low that the asteroid cannot hold on to an atmosphere. Conversely, black holes have escape velocities that exceed the speed of light.

EVENT HORIZON: Area surrounding a black hole that is defined by the Schwarzschild radius. Light emitted from within this area cannot escape from the gravitational pull of the black hole.

EXPANDING UNIVERSE: Movement, measured by the Doppler effect, of objects in the universe away from each other, proving that the universe is expanding. Distant objects recede more quickly than nearby ones.

FIREBALL: Bright meteor whose magnitude is at least −4. Much larger than normal meteors, some fireballs are bright enough to rival the sun.

GALAXY: System of billions of gravitationally connected stars. Galaxies generally form clusters, most of which are further grouped into clusters of clusters, or superclusters. Elliptical galaxies may arise from collisions between spiral galaxies.

GEOSTATIONARY (GEOSYNCHRONOUS) ORBIT: Circular orbit that is 22,300 miles (35,900 kilometers) above the equator and whose speed therefore matches that of the Earth's rotation, 1,038 miles per hour. Viewed from Earth, a satellite at this orbit appears to be stationary. Useful for communications and weather satellites.

HUBBLE SPACE TELESCOPE (HST): Orbiting telescope. After its launch on April 25, 1990, a "spherical aberration" flaw in the main mirror was discovered, resulting in unsharp images. However, during an *Endeavour* mission in December 1993, this and other problems were fully corrected. The HST now performs to original specifications and delivers the clearest telescopic images of any instrument yet constructed.

IAU: International Astronomical Union.

INTERNATIONAL ASTRONOMICAL UNION (IAU): Organization founded in 1919 that now includes thousands of astronomers from around the world.

JET PROPULSION LABORATORY (JPL): Branch of NASA located in Pasadena, California, its Space Flight Operations Facility site, the mission control center for robotic space and moon probes. JPL is famous for its processed images from the Voyager, Magellan, and other missions.

K: Symbol for degrees Kelvin, the unit for temperature measured from absolute zero (−273.15 degrees Celsius; −459.67 degrees Fahrenheit). A Kelvin degree is equal to a Celsius degree; water freezes on the Kelvin scale at 273.15 degrees, or 0 degrees Celsius.

KECK TELESCOPE: Either of two identical 400-inch (10-meter) reflecting telescopes on Mauna Kea, Hawaii. As of 1998, they are the world's largest telescopes.

LAGRANGIAN POINTS: Five points in which a small body can remain in equilibrium in the company of two massive bodies that are in mutual orbit around each other. Only two of these points are stable. They are named after Joseph-Louis Lagrange (1736–1813), French astronomer and mathematician.

LASER: Acronym for light amplified by standard emission of radiation. Lasers function because atoms can absorb only a certain amount of energy before their electrons move to a higher energy level. When a photon of just the right energy shines on an atom in an excited state, it stimulates the atom to emit an identical photon. This second photon moves in the same direction as the first, and with equal energy. Energy pumped into a laser-making device continually produces atoms at a higher energy state; mirrors are then used to reflect the photons back and forth until the light becomes so intense that it escapes through a partially reflecting mirror as a beam whose light is all of the same frequency and phase.

LEONID METEORS: Annual meteor shower appearing around November 17. This shower is associated with Comet Temple-Tuttle, with the greatest concentrations occurring every thirty-three years, when the comet passes closest to Earth.

LIGHT: Portion of the electromagnetic spectrum to which the human eye is sensitive.

LIGHT, VELOCITY OF: Speed light travels in a vacuum (sometimes represented by the letter *c*). Equal to 186,282.397 miles (299,728.377 kilometers) per second. Light travels more slowly through denser media such as water or glass.

LIGHT POLLUTION: Light from cities and other sources that brightens the sky and interferes with astronomical observing.

LIGHT-YEAR: Distance light travels through a vacuum in a year, equal to 5.8786 trillion miles (9.4587 trillion kilometers). It is a measure of distance, not time.

LOCAL GROUP: Group of about thirty galaxies, including the Milky Way, held together by mutual gravitation. The largest galaxy in the Local Group is the Andromeda Galaxy, whose mass is about twice that of the Milky Way.

MAGELLANIC CLOUDS: Two small, irregular galaxies orbiting the Milky Way at a distance of 170,000 light-years. They were named after Ferdinand Magellan, the first prominent European to describe them. They are visible to the naked eye from around the equator and points southward.

MAGNITUDE: Measure of an object's brightness instituted by Hipparchus (146–127 B.C.). His classification system had six levels, with one being the brightest stars and six comprising stars barely visible to the unaided eye. First magnitude stars are 100 times brighter than sixth magnitude stars, making each magnitude 2.5 times brighter than the previous one.

METEOR: Extraterrestrial particle, commonly the size of an apple seed, that slams into Earth's atmosphere and often produces a bright trail. Meteors tend to follow in the wake of a comet. Six meteors can be seen per hour on any clear night, more during meteor showers.

METEORITE: Material of a meteor that hits the Earth after surviving its passage through the atmosphere.

METEOROID: Interplanetary object smaller than an asteroid. It is called a meteor if it enters Earth's atmosphere and termed a meteorite in the unlikely event it survives all the way to the ground.

MILKY WAY: Spiral galaxy composed of 400 billion stars, plus or minus 50 percent, and home to our sun and planets. The Milky Way is disk-shaped, with a large central bulge (nucleus) and is 100,000 light-years in diameter.

MOON: Earth's only natural satellite. The moon is 4.6 billion years old, has a mass of 7.4×10^{22} kilograms, is 2,157.6 miles (3,480 kilometers) wide, and orbits the Earth at an average center-to-center distance of 238,329.24 miles (383,472 kilometers).

MOUNT PALOMAR: Site of Palomar Observatory's 200-inch (508-centimeter) reflecting Hale telescope, the world's largest for a quarter-century. Located outside of San Diego, it is more properly called Palomar Mountain.

NASA: National Aeronautics and Space Administration. Established in July of 1958, NASA is a federal agency authorized to regulate all civilian aeronautic research and space programs.

NEBULA: Region of dust and gas in space, such as the Orion Nebula or the Lagoon Nebula. Some nebulae glow by themselves and are usually red; others (for example, the gas enveloping the Pleiades) reflect the light of nearby stars and are blue. Absorption nebulae appear as a dark pattern against a light background. Nebulae are places where starbirth can occur.

NEUTRON: One of the basic constituents of an atom, always located in the nucleus. Neutrons are slightly heavier than protons but carry no charge.

NEUTRON STAR: Star so dense that the electrons and protons have been forced together to form an ocean of neutrons. Neutron stars can be formed by supernova explosions in which the collapse of the star is halted by rising neutron pressure. A neutron star has a mass several times greater than that of the sun but a diameter of only a few miles.

NGC: *New General Catalogue of Nebulae and Clusters of Stars*. This catalog, first compiled in 1888, is now the standard list of all deep-sky objects.

NOVA: Explosion on the surface of a star. It can be caused by the addition of a large amount of matter, which causes the star to become a hundred thousand times as luminous as normal. Novas are named after the constellation in which they are found and the year in which they are first seen. Often the star was invisible before the explosion.

OLBERS' PARADOX: Surprising answer to the puzzle of why the night sky is dark. Nineteenth-century astronomer Heinrich Olbers (1758–

1840) noticed that if one assumed that the universe is infinitely old and contains a vast number of stars, every line of sight should lead to the surface of a star, causing the night to be brilliant. Various reasons have been proposed for the obvious fact that the night is dark. Olbers thought that the reason was blockage by interstellar dust; more recently the cause was ascribed to the red-shifting of all the receding galaxies. Now the usual explanation is that the universe is still young, thus, starlight has not had time to reach us or to fill the universe with light.

OORT CLOUD: A spherical cloud of icy, tailless comets that surrounds the sun at a distance of about 1 light-year, first suggested by Jan Hendrik Oort (1900–1992) in 1950. When perturbed by passing stars, the Oort cloud hurls comets toward the solar system. This explains the source by which comets are replenished.

OPPOSITION: Period when a superior planet arrives at its minimum distance to Earth and stands directly opposite the sun in our sky.

OSCILLATING UNIVERSE: Cosmological model contending that the universe has cycles of expansion and contraction. If there is enough matter in the universe, deemed unlikely at present, then the current expansion will be halted and the universe will slowly begin to collapse. Some astronomers and philosophers surmise that after such a "big crunch," a rebound might cause a new universe to be formed.

PARSEC: Abbreviation for *par*allax of one *sec*ond, astrophysicists' preferred distance unit, the distance at which a body would display an annual parallax of one arc second as Earth orbits the sun. A parsec is equal to 3.2616 light-years. A kiloparsec is 1,000 parsecs, a megaparsec is 1 million parsecs.

PERIGEE: From the Greek words meaning "near Earth," the point in an artificial satellite's (or the moon's) orbit when it is nearest to Earth.

PERIHELION: From the Greek words meaning "near sun," the point at which an object in a noncircular orbit around the sun is nearest to the sun. Opposite to APHELION.

PERSEIDS: Major annual meteor shower that peaks on August 11 or 12.

PHOTOMETRY: Science of measuring an object's brightness.

PHOTOSPHERE: Gaseous surface of the sun. Its relatively dense gas is 200 miles (320 kilometers) thick, at a temperature of 5,800 degrees Celsius (11,000 degrees Fahrenheit). Granules, which reveal the sun's convection processes, are visible on the photosphere through properly equipped small telescopes.

PLASMA: Ionized gas consisting of pieces of atoms. As gas is heated, collisions with other atoms or strong radiation knock electrons out of their shells, leaving a soup of electrons and naked nuclei. Most of the visible matter in the universe is plasma.

POLE STAR: Popular name for Polaris. However, because of the twenty-six-millennium wobble in Earth's axis, called precession, Alpha Cephei will be the polestar in about five thousand years.

PULSAR: Source of radio waves with short, extremely regular repeating patterns. Pulsars are spinning neutron stars that sometimes sit at the center of supernova remnants. More than one hundred pulsars are now known; the youngest have periods of as short as 860 rotations per second. A few have been observed in visible or X-ray light.

QSO: *Q*uasi-*s*tellar *o*bject, or quasar.

QUASAR: Starlike source of radio emission. About two hundred quasars have been identified, each emitting the energy of one hundred normal galaxies. Since quasars vary their brightness over a few hours, they cannot be much bigger than our solar system. Quasars appear to be the cores of young galaxies that emitted fantastic amounts of energy when the universe was young.

RADIO TELESCOPE: One or more antennae, usually parabolic, used to receive celestial radio signals. Because radio waves have wavelengths far longer than visible light, much larger telescopes must be used to obtain comparable resolution.

RED DWARF: Smallest normal star and probably the most common type of sun. Eighty percent of all stars are red dwarfs. Red dwarfs have a cool surface and, since they fuse their nuclear fuel frugally, can continue to shine for at least 30 billion years, but with a luminosity just 5 percent to .01 percent that of the sun. For this reason, red dwarfs can be detected only to a distance of about 120 light-years.

RED GIANT: One of the last stages in a star's life cycle, when its outer layers swell to ten to one hundred times the sun's diameter. These stars cool to a surface temperature of 2,000 to 3,000 degrees Celsius (3,600 to 5,000 degrees Fahrenheit) and shine as brightly as one hundred suns only because they have so much surface from which to radiate. Betelgeuse and Antares are examples.

RED SHIFT: Lengthening of a wavelength of light due to an object receding from an observer. Cosmologically, red shifts indicate that the universe is expanding: The farther an object, the faster it recedes. The currently accepted value of expansion is expressed as 50 to 75 kilometers per second per megaparsec of distance from us, or about 10 miles per second per million light-years.

RED SPOT: Enduring hurricanelike storm located 24 degrees south of Jupiter's equator, first observed by Giovanni Cassini (1625–1712) in 1666. An elliptical cyclone with a width of 22,000 miles (35,000 kilometers), it's Jupiter's only continuous atmospheric feature, and is caused by powerful currents produced by the different speeds of rotation of adjacent bands of Jupiter's cloudy atmosphere.

RELATIVITY: Amply proven theory describing the unity of time and space, and motion and gravity, proposed by Albert Einstein.

The *special theory of relativity*, published in 1905, postulated that every observer will measure the same value for the speed of light. It stated that two observers, one stationary and one moving, will see the same event differently. Following from this is the idea that time slows down, length contracts, and mass increases at higher speeds and in stronger gravitational fields.

Einstein's *general theory of relativity*, published in 1915, describes how mass distorts space and time, and how gravity and acceleration are the same. Just as a cannonball on a trampoline depresses the fabric, any mass stretches space and time. General relativity predicted that light must traverse these curved lines, a theory that was confirmed in 1919 when, during a total eclipse, starlight grazing the sun was displaced exactly as relativity predicted.

RESOLUTION: Telescope's ability to distinguish two close objects or fine detail. Resolution limitations of Earth-bound telescopes mostly come from the atmosphere, which blurs images.

ROCHE LIMIT: Smallest distance at which a natural satellite can orbit without being destroyed by tidal forces. For a planet and moon of similar composition, this distance is 2.5 planet radii. The moon Io sits just outside Jupiter's Roche limit. Saturn's rings reside within its Roche limit. The Roche limit is named for the French astronomer Edouard Roche (1820–1883), who calculated its existence in 1848.

SAROS: Cycle of 6,585.3 days (18 years, 11.3 days) during which the sun, moon, and Earth return to the same relative positions, causing the same kind of eclipse to repeat itself.

SATELLITE: Any smaller body orbiting a larger body, although with two nearly equal-sized masses the difference can be blurry, since in reality two objects always merely orbit their common center of mass. Satellites can be natural, like the moon, or artificial. The first artificial satellite, launched October 4, 1957, was the Soviet craft *Sputnik I*. The first American satellite was *Explorer I*, sent aloft on January 31, 1958.

SCHWARZSCHILD RADIUS: Radius of the event horizon in a black hole. When light or matter passes the Schwarzschild radius going inward, it cannot escape the gravitational pull of the dense body because, within this radius, the escape velocity exceeds the speed of light.

The sun is not massive enough to have sufficient gravity to collapse into a black hole, but if it could, its Schwarzschild radius would be only about 2 miles (3 kilometers), while the Earth would be only about 1 inch (3 centimeters) in radius, if turned into a black hole.

SETI: Search for Extraterrestrial Intelligence. This project, privately funded, uses radio telescopes to look for signals (deliberate or inadvertent) from alien civilizations. The main computer analyzes 8 million radio frequencies, searching for signals that cannot be of natural origin.

SHOOTING STAR: Common name for a meteor.

SINGULARITY: Matter crushed into a point with zero volume. All of the mass of a black hole is concentrated at this point, forcing space and time to curve back on itself. Also, at one time, before the Big Bang, all the mass in the universe was contained in a singularity. The known laws of physics do not apply at a singularity.

SOLAR CONSTANT: Amount of heat striking a surface perpendicular to the sun's rays just outside the Earth's atmosphere. The accepted value is 2 calories per minute per square centimeter. This equals about 1.3 kilowatts per square meter.

SOLAR CYCLE OR SUNSPOT CYCLE: Uneven eleven-year period in which sunspot activity repeats its maximum and minimum values. The sun's magnetic field flips direction every sunspot cycle and thus displays a twenty-two-year rhythm. Generally, the more spots, the warmer our planet becomes.

SOLAR SYSTEM: The sun and its family of orbiting planets, asteroids, and comets.

SOLAR WIND: Continuous gust of atomic particles radiating from the sun's corona. Mostly protons and electrons with a few helium nuclei, the solar wind moves outward at 220 to 500 miles (350 to 800 kilometers) per second and require four days to reach Earth. The solar wind averages five protons and five electrons per cubic centimeter as it rushes past our planet. The solar wind blows back the tails of comets, causing them always to point away from the sun. Solar wind also sweeps the Earth's magnetosphere into a cometary shape and induces the Van Allen radiation belts. Erupting solar flares intensify the solar wind, allowing its particles to follow Earth's magnetic-field lines downward into the atmosphere, causing auroras.

SPACE SHUTTLE: Reusable manned spacecraft. NASA's space shuttle can haul 65,000 pounds (29,500 kilograms) of equipment into orbit and return 35,000 pounds (16,000 kilograms) back to Earth. At launch, the shuttle relies on twin solid-fuel boosters that flank a huge external fuel tank carrying liquid hydrogen and oxygen to power the orbiter's three main engines.

SPECTRAL LINES: Narrow lines seen when an object's light is spread into its component wavelengths though a spectroscope or spectrograph.

In 1814, Joseph Fraunhofer (1787–1826), studying the sun's spectrum, identified many elements found on Earth and observed the lines of an unknown element, which he named helium, from *helios* (Greek for "sun"). Every element emits its own characteristic spectral lines.

SPECTROSCOPE OR SPECTROGRAPH: Instrument used to observe the spectra of stars or galaxies. It uses a diffraction grating to separate light by wavelength.

SPECTRUM: Entire range of electromagnetic radiation with wavelengths spanning gamma rays through radio waves to microwaves. Sometimes used to refer to just the visible portion, the rainbow colors.

SPIRAL GALAXY: Type of galaxy with arms winding away from the center. Some, called barred spirals, have just two arms snaking away from the ball-shaped nucleus. However, most spiral galaxies look like pinwheels and show bands of stars and dust twisting away from their core.

STAR: Enormous sphere of hot gas that, at some point in its life, fuses hydrogen into helium. The sun is an example of a typical star. Under a moonless sky, away from streetlights, about three thousand stars appear to the naked eye. A telescope reveals millions more.

SUNSPOT: Relatively cool, dark patch on the surface of the sun. Sunspots range from blotches 900 miles (1,500 kilometers) wide to systems spanning 100,000 miles (161,000 kilometers). As with other solar magnetic phenomena, sunspot frequency follows a somewhat irregular eleven-year cycle. Sunspots rarely venture farther than 40 degrees away from the sun's equator or closer than 10 degrees to it. Most groups last about ten days.

SUPERCLUSTER: Collection of clusters of galaxies, the largest structure in the universe. These enormous assemblies resemble sponges, with galaxies forming curving sheets enclosing vast voids. Our Local Group is part of the Virgo supercluster. The average supercluster has twelve galactic clusters (each of which in turn may contain thousands of galaxies) and spans hundreds of millions of light-years.

SUPERGIANT STAR: Biggest and brightest star. Supergiants contain from twenty to several hundred times the mass of the sun and shine a million times more brightly. We see relatively few of them because supergiants live just a few million years and, additionally, only about one star in ten thousand is a supergiant. However, these powerhouses shine across great distances; when one views spiral arms of distant galaxies, only the light of supergiant stars are visible.

SUPERIOR PLANET: Planet that orbits farther from the sun than Earth. All the planets from Mars to Pluto are superior planets.

SUPERNOVA: Titanic stellar explosion.

Type I: In a double-star system in which one star is a white dwarf and the other a red giant, enough material from the red giant may fall onto the dense white dwarf to allow it to ignite into a supernova explosion.

Type II: Toward the end of their brief lives, massive stars lose their ability to produce nuclear energy. When the star can no longer support its great weight by outward radiation pressure, its core collapses in an instant to form a sphere of neutrons. This state stabilizes the core, and the star rebounds like a trampoline, generating a shock wave and a burst of neutrinos that rip through the star and detonate the greatest explosion in the known universe.

For a few weeks, supernovas shine as brightly as an entire galaxy, allowing them to be seen billions of light-years away. These titanic explosions create all elements heavier than iron. Thus, in a real sense, humans are made of stardust; every heavy element in our bodies, such as the iodine in our thyroid glands, once churned in the fires of a supernova.

After the supernova, the star's outer layers blast into the surrounding space, creating a nebula that grows for millennia. Supernova shock waves ripple through the arms of spiral galaxies. They enrich the interstellar medium with heavy elements and perturb nebulae, which coalesce into new stars. They also may produce the cosmic rays streaking throughout the universe. There may be one supernova in each galaxy per century, on average.

SUPERNOVA 1987A: Closest and brightest supernova observed from Earth since 1604; first seen in the Large Magellanic Cloud on the night of February 24, 1987. Its progenitor star was Sanduleak – 69° 202. At its peak in mid May, it reached magnitude 2.8.

SYNCHRONOUS ROTATION: Situation that exists when the period of a celestial body's rotation and its period of revolution around its parent body are the same. The moon, for example, always displays the same face to Earth because of its synchronous rotation. In the distant future, tidal drag will force the Earth into a synchronous rotation, always displaying the same face to the moon.

TUNGUSKA EVENT: Tremendous explosion observed in Siberia on June 30, 1908, estimated to occur 5 miles (8 kilometers) above Earth. The source was either a meteor or a comet.

UFO: *U*nidentified *f*lying *o*bject. Although UFOs are often reported by the public, the great majority prove to be aircraft, bright stars or planets, meteors, satellites, etc. The rest of the sightings are unexplained. Astronomers do not believe UFOs are extraterrestrial visitors, and astronomers never seem to see them.

UNIVERSE: Everything there is. The universe contains all space, time, energy, and matter. However, some use the term *universe* to mean everything that can ever be seen rather than everything that exists. If there are other realms that, for various reasons, must remain forever hidden from view, it is now common to call them *other universes*.

VARIABLE STAR: Star that varies its brightness in a regular or irregular cycle. About one third of all stars alter their light for various reasons.

VOYAGER: Pair of identical robot craft, the most successful probes to date, launched in 1977 to investigate the Jovian planets. NASA exploited an alignment of the outer solar system to use gravity assists to slingshot past each planet onward to the next destination. While *Voyager 1* visited only Jupiter and Saturn, *Voyager 2* explored Jupiter in 1979, Saturn in 1981, Uranus in 1986, and Neptune in 1989. The Voyagers returned tens of thousands of photos; during their brief encounters they revealed more about the outer planets than astronomers had learned in three hundred years of telescopic observation.

WHITE DWARF: Old star whose diameter and luminosity are respectively one hundredth and one ten-thousandth less than those of the sun, but whose density is 1 million times greater than that of water. The white-dwarf stage is a common one in most stars, those with less than 1.4 times the mass of the sun. When these stars have exhausted a critical amount of hydrogen in their core, they start to collapse. After some intermediate stages, they no longer produce energy from fusion, and their light comes entirely from the escape of their internal heat.

ZODIAC: Band 15 degrees wide, circling the sky, through which the moon, the planets (excluding Pluto), and the sun seem to travel. In addition to the classical zodiacal constellations (Aries, Taurus,

Gemini, Cancer, Leo, Virgo, Libra, Scorpius, Sagittarius, Capricornus, Aquarius, and Pisces), solar-system objects also pass through Ophiuchus. Because all but five of these constellations represent animals, the band was named *zōidiakos kyklos* ("circle of animals") by the Greeks.